They identify habitat loss, degradation, fragmentation, the introduction of non-native species, and over harvesting as metacauses of a new rate of extinction. While a conservative estimate of the current rate of species loss is 27,000 species per year, far more species are considered endangered, vulnerable, or rare (Wilson 1992:280). Conservation biologists use the language of apocalypse, hemorrhaging, and holocaust to describe the decline in the abundance and variety of life forms.[7]

The emergence of biodiversity as a new form has also coincided with the global rise of the nongovernmental organization (NGO). NGOs are non-state institutions that nevertheless affect policy and aim to transform debates across national borders. Keck and Sikkink (1998) have described the influence of what they call "transnational advocacy networks." Activists forming these transnational networks (scientists along with women's, labor, and human rights advocates) are motivated primarily by values rather than economic gain. Networks of scientists and others concerned with the value of biodiversity and its loss link activists across borders; the World Wide Fund for Nature (WWF), Conservation International (CI), and The Nature Conservancy (TNC) are three of the largest advocacy NGOs forming the institutional backbone of this transnational assemblage.

Further, biodiversity has emerged at a contingent moment in postcolonial history. Under the logics of natural history and wildlife conservation (at least until the early 1990s), EuroAmerican experts held the important positions of authority in scientific nature-making projects across the tropical world.[8] This condition is increasingly rare, however. For example, when I first began working in Indonesia in 1994, the Jakarta offices of WWF, CI, and TNC all had EuroAmerican administrators, while by the time I left in 1997 each organization had hired an Indonesian director to oversee its domestic programs. Positions of leadership and authority in field biology and conservation management are now occupied by scientists and other experts from the South. The conjuncture of biodiversity's tropical geography, the emergence of indigenous scientific expertise, and increased assertions of domestic bureaucratic authority in the realm of nature conservation, have shaped the particular understanding of biodiversity that this book will explore in detail.

Biodiversity conservation in the 1990s often proposed a particular solution to the problem of nature and the human in the form of the Integrated Conservation and Development Program (ICDP). Noting the ubiquity of conservation failures, the ICDP was premised on an understanding that previous efforts at wildlife conservation had not taken into sufficient consideration the needs of the people who live around conservation areas. These needs were interpreted in rational economic terms (by the biologists and economists who environmental NGOs tend to employ) as the ability

to derive income from surrounding natural areas. If alternative income sources could be found, the theory went, then people would stop hunting, fishing, gathering, felling, burning, planting, and all the other activities that threaten rare plants and animals in and around protected areas. Increased access to markets, land privatization, and ecotourism were key components of this neoliberal solution.

As a cultural formation, biodiversity conservation can be tracked globally. In order to understand conservation in a generative way, however, rather than as a set of established discourses, it is necessary to graph it at its specific sites of production. *Wild Profusion* is an ethnographic account of the rationalities surrounding a particular instance of mid-1990s biodiversity conservation. It concerns how biodiverse nature was made in the Togean Islands of Sulawesi, Indonesia between 1988 and 1998, how the main actors in the Togean conservation project (Indonesian biologists and Togean people) constituted and were constituted through projects of nature-making, and how the nation was critical to both the particularity of the Togean biodiversity project itself and to the subjectivities formed within the context of Indonesian science. Examining the ensuing configuration in a specific locality allows us to understand the emergent rationalities and identities, and the multiple natures, resulting from the project to conserve Togean biodiversity.

The Togean Islands first appeared as a potential conservation area in the early 1980s. Following upon traces in the scientific literature written by late-nineteenth- and early-twentieth-century natural historians who had documented bird and coral varieties there, Indonesian and EuroAmerican scientists arrived to establish the potential of the site for a nature reserve.[9] Then in the early 1990s, through the work of conservation biologist Jatna Supriatna and his students, the Togean research station, Camp Uemata, was built and an enduring project of conservation research and management commenced. Two institutions, the new Indonesian Foundation for the Advancement of Biological Sciences (IFABS) based in Jakarta, and Conservation International based in Washington, D.C., then jointly set their sites on turning the islands into a national park.[10]

I call this work a multisited ethnography in a single locality because of what the Togean Islands as a singular site can reveal of the travels of cultural meanings, objects, and identities across wider fields of engagement. Although biodiversity was a transnational practice, it took on shape and specificity through the work of Indonesian biologists to document species and implement a program of conservation and development, and the islands and their biophysical properties always meant different things to different people. The project collapses easy definitions of "nature" since Indonesian scientists, EuroAmerican biologists, commercial traders,

bureaucrats, and diverse Togean people each engaged with Togean land and marinescapes in discontinuous ways—producing the archipelago as contrastive and contested "sites." Rather than the "conventional *mise-en-scène* of ethnographic research" (Marcus 1995), the Togean Islands should be understood as a locality generative of cosmopolitan imaginings of science, nation, and biodiversity conservation.

The Togean archipelago is positioned particularly well to tell this story, located, as it is, at the intersection of three significant lines: Wallace's Line; the equator; and the tip of a strand of rattan that, legend has it, connects the islands to the former Sultanates of Ternate and Bone. These are the threads to which we will now turn.

Wallace's Line

In the mid-nineteenth century, Alfred Russel Wallace, the naturalist who devised a theory of natural selection independently from and simultaneously with Charles Darwin, observed a division in the morphology of birds and mammals across a line that separates the islands of Bali and Lombok in the south, and Borneo from Sulawesi in the north. "Wallace's Line," as it became known, divided the eastern half of the Indonesian archipelago from its western portion and demarcated a biogeographic division between Asiatic and Australian fauna. Later marks on the map (a second Wallace's Line, Lydekker's Line, Huxley's Line, and Weber's Line) express the controversy that once existed over precisely where the proper division was to be made (Daws and Fujita 1999:74). Now, the region encompassing the island of Sulawesi is understood to be a zone of transition and is referred to by the name "Wallacea." In general terms, to the Australian side of the zone the marsupial order dominates among the mammals, while the Asiatic side is dominated by placental mammals.

In the collision between Asian and Australian land masses, the fauna of these regions became mixed and commenced a unique pattern of biological evolution (Whitten, Mustafa, and Henderson 1987:37–52). Sulawesi's separation from Borneo by a deep ocean trench, and the complexity of its geologic history, have given the island a very high level of vertebrate endemism: 62 percent of Sulawesi's 122 mammals are endemic. If one were to eliminate bats from the calculation, the rate is closer to 98 percent, striking compared with the 18 percent rate of endemism on the neighboring island of Borneo. More species of macaque monkey exist on Sulawesi than anywhere else in the world, and of more than 300 species of birds, 30 percent are endemic, the highest figure for any island other than New Guinea. Sulawesi is home to the Maleo, a bird that incubates a 250 gram egg in the island's volcanic sands, and the Babirusa, a "deer

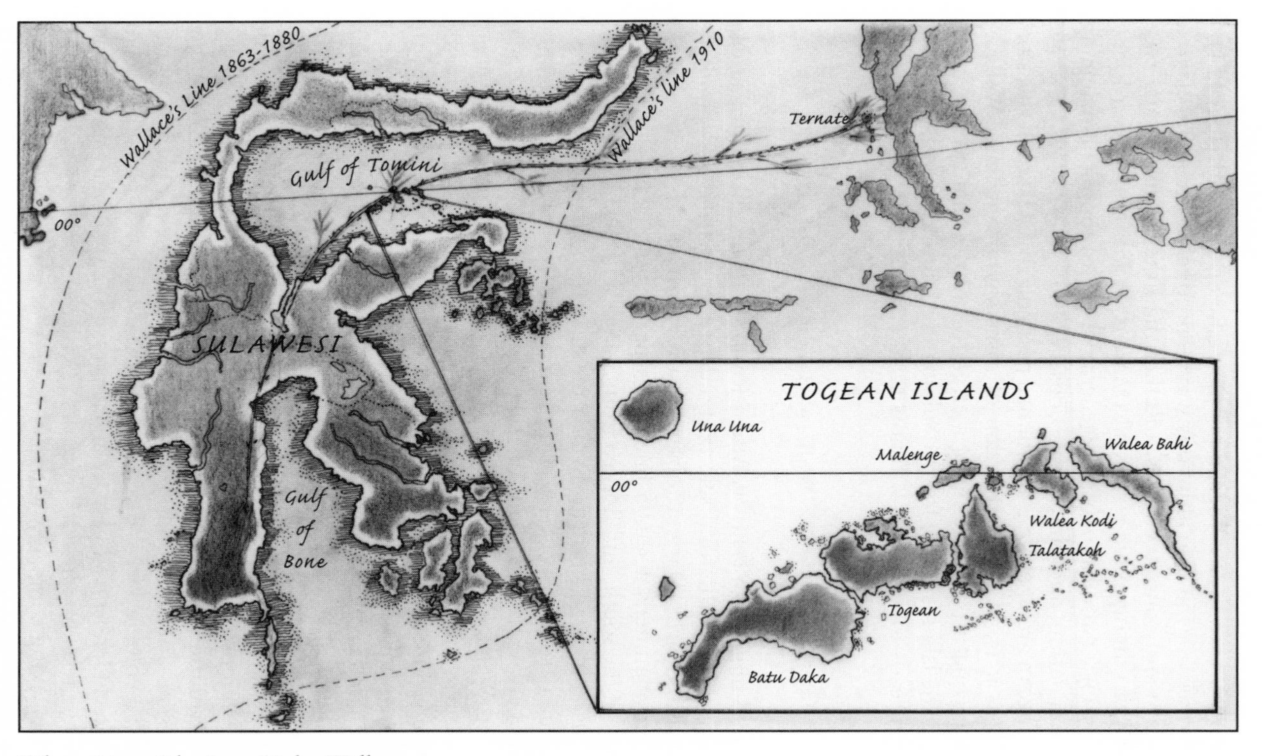

"Three Lines," by Jane Bixby Weller

pig" with four long tusks that curl back toward its head. The largest Sulawesi mammal is the Anoa, a dwarf genus of buffalo of which there are two species. Two kinds of cuscus, the Bear cuscus and the nocturnal Tree cuscus, and the world's largest snake, the Reticulated python, live in Sulawesi trees. And the island has four high lakes, each with its own endemic fauna, including an unusual species of blind shrimp only found in caves.

What to make of this profusion of unusual and fascinating creatures? On the one hand we might insist, along with many biologists, that the meaning and value of this abundant nature is self-evident, the importance of such unusual and natural diversity speaks for itself. This perspective is intrinsic in the writings of biologist Edward O. Wilson, who describes this way his passion for enumerating the "diversity of life":

> In the realm of physical measurement, evolutionary biology is far behind the rest of the natural sciences. Certain numbers are crucial to our ordinary understanding of the universe. What is the mean diameter of the earth? It is 12,742 kilometers (7,913 miles). How many stars are there in the Milky Way, an ordinary galaxy? Approximately 10^{11}, 100 billion. How many genes are there in a small virus? There are 10 (in a ϕX174 phage). What is the mass of an electron? It is 9.1×10^{-28} grams. And how many species of organisms are there on earth? We don't know, not even to the nearest order of magnitude. (Wilson 1992:132)

On the other hand, anthropologists, historians, and science studies scholars who do not claim the value of nature as singular or self-evident, have been at the center of debates over the social constructedness of sex and gender, kinship, race, and other formulations that might otherwise be articulated in biological terms. In relation to the biophysical environment, social scientists have argued that people are involved in shaping many spaces that are overdetermined to be "natural," that the idealist separation of "humans" from "nature" is an event historically and culturally specific to the European Enlightenment, and that any understanding of nature will always depend upon processes of representation and the subjectivity of those claiming to represent such a nature.

Along similar lines, science studies scholars are responsible for intriguing efforts to get beyond the sense that nature must be *either* foundational or found only in language. In the wake of the so-named "science wars," this group of thinkers has made intense efforts to "take nature seriously," without necessarily taking its forms as self-evident.[11] For example, Andrew Pickering, who has studied the particle physics of Donald Glaser, writes, "Now I can clarify my sense of material agency. It is simply the sense that Glaser's detectors *did* things—boiling explosively or along the

lines of tracks or whatever—and that these doings were importantly separate from Glaser. To understand what happened when Glaser took a passive role in the dance of [material] agency, I think one has to acknowledge that some other source of activity and agency was at work, and in this case that source was the material world" (Pickering 1995:51–52). Yet, material agency is not the same thing as intentionality, Pickering argues: "My argument is that we need to recognize that material agency is irreducible to human agency if we are to understand scientific practice. Nevertheless, I need to stress that the trajectory of emergence of material agency is bound up with that of human agency. Material agency does not force itself upon scientists" (53).

Likewise, Bruno Latour has argued that "nature" is an ally of "truth" only when all scientific controversies are settled. Before that moment, no one really knows what nature will say. "When you attack a colleague's claim, criticize a world-view, modalize a statement you cannot *just* say that Nature is with you: 'just' will never be enough. You are bound to use other allies besides Nature. If you succeed, then Nature will be enough and all other allies and resources will be made redundant" (Latour 1987:98). This is not to argue, he would say, against the existence of substances and objects in historical time, yet we have to understand such historicity in terms of the *production* of things: "History of science does not document the travel *through* time of an already existing *substance*. Such a move would accept too much of what the giants demand. Science studies documents the modifications of the ingredients that compose an articulation of entities" (Latour 1999:162).

These scholars of technoscience help us to see how an object becomes a different "thing" or "substance" in the world over time, with each new "articulation of entities" or set of associations encompassing it. An attention to forms, assemblages, conjunctures, and histories will help us to understand how "Togean nature" would have been a different object in Wallace's period of ninteenth-century natural history (where, for one, specimens should be reposited in museums by colonial governments to teach us earth's history) than in Indonesian scientists' biodiversity (where, instead, nature must be preserved in situ for some combination of "human" and "Indonesian" posterity). Emphasizing the active materiality of nature helps us to keep in focus both the unusualness of faunal forms in the region called "Wallacea," and the moment in time when scientists reassembled matter, institutions, experiments, and identities around Togean Island plants and animals to produce a nature that could be named "biodiverse." It also lets us predict that the green and blue hues of Togean land and marinescapes may be something altogether different for Togean people who assembled other material, institutional,

experimental, and identitarian forms around an encompassing Togean flora and fauna.

Hemispheric Divides

A second line running across the Togean archipelago is the equator. The equator marks a metageography (Lewis and Wigen 1997) dividing global "North" from global "South." This invisible line is a displacement of the earlier metageography, "East" and "West," which for many generations has represented the specter of deep inherent divisions—divisions of science, reason, modernity, development, and similar civilizational ruptures. From the perspective of such a divided world, the term "Indonesians' science" would be an oxymoron, since the global South purportedly stands for belief and unreason. From a more critical perspective, however, "Indonesians' science" demonstrates the limitations, indeed the falsehood, of this modernist parsing of North and South. By now, the question is not whether the divide is accurate or not, but rather what the "truth-effects" of such a proposition are. What kind of labor does the "Western science" synecdoche perform in the process of creating identities and expectations? The truth-effects of hemispheric divides are a problem that will thread their way continually across the narrative of this text.

How do we study these effects in relation to conservation biology in Indonesia? In moving past the terrain of ethnoscience (disavowing its tendency to bifurcate types of knowers), in rethinking the idea of the field (instilling a focus on elite subjectivity), and in provincializing EuroAmerican science itself (one response to anthropology's autocritique), anthropologists and other science studiers now take as their problem spaces the laboratory, the nuclear test site, cyberspace, and the hospital setting. To the extent that we have reincarcerated science within the bounds of Europe and North America in the process, science remains a metonym for EuroAmerican modernity and rationality, and Indonesia and the rest of the global South remain the lands of farmers and fishers living out "alternative" modernities. But what of science and reason produced beyond EuroAmerica? What of the natural sciences in Indonesia that do not fit into those locally circumscribed forms of knowledge conventionally studied as "ethnoscience" or "indigenous knowledge"?

When I first began to search for a research site in eastern Indonesia in 1994, I visited three different conservation projects: Bunaken National Park in North Sulawesi, Taka Bonerate Park in South Sulawesi, and the Togean Islands. Both Bunaken and Taka Bonerate were managed by EuroAmerican consultants employing Indonesian field and office assistants,

and only the Togean Island project was run entirely by Indonesian scientists and staff. I believed (it turned out, correctly) that I would learn more interesting things from these Indonesian experts than I could from expatriate consultants with whom I shared a quite similar educational and social history. Although I could not predict what I would find, I assumed that there would be something "different" about the Togean project as a result of its nationalization.

My research with the scientists entailed living and working at Camp Uemata, participating in the scientists' ICDP activities in the Togean Islands, attending conservation meetings with IFABS and CI in Jakarta and Palu (the capital of Central Sulawesi), and interviewing Indonesian and EuroAmerican conservation practitioners in Jakarta, Manado, and Central Sulawesi. What I discovered in the process of accompanying biologists in their species inventories, or participating in meetings and discussions about Togean conservation, was that difference and similarity are both part of the story of biodiversity in Indonesia. Indonesians' scientific and conservation management practices were distinct from, but also very much the same as, what could be found elsewhere in the global travels of the mid-1990s biodiversity paradigm.[12] I came to understand that these questions of comparability were also important to Indonesian scientists who sought recognition but also desired a sphere of autonomy for their work.

I witnessed, for example, the frustration Indonesian scientists sometimes felt in their collaborations with foreign scientists, and heard the comments some of these same EuroAmerican visitors made. "The only thing disappointing about this experience at the [Togean] research station," one foreign scientist said, "is that there are no *real* scientists working here." Processes of negation were constitutive of the Indonesian subject of 1990s conservation biology. While science studies is an invaluable rubric through which to understand Indonesians' conservation biology, it is also insufficient to explain the interplay of similarity and difference that exists across the imagined hemispheric divide. The "postcolonial condition," as it is known, goes further in explaining the subjectivities produced through Togean science where Indonesians struggled to attain a voice in an international field.

Although the autonomy of the nation has secured a space for Indonesians within transnational science, and these biologists are now mentors, partners, and colleagues of EuroAmerican scientists in producing natural scientific knowledge, they still face many challenges: the terms for what can be considered "good" science are often set somewhere else; Indonesians are frequently expected to contribute only data, rather than theory; Indonesian scholars are required to know EuroAmerican literatures while the inverse is not usually the case; and EuroAmerican scientists continue

Photographing biodiversity, by Celia Lowe.

to take for granted that Indonesia is only one "problem space" within an entire world amenable to their investigations, while Indonesian scholars tend to perceive the nation as their most pressing area of concern. For these reasons, and more, Indonesian scientists struggle for recognition within transnational scientific domains.

Postcolonial theorists have contributed greatly to our understanding of this dilemma. Itty Abraham (1998), Partha Chatterjee (1993), Pheng Cheah (2003), Deepak Kumar (1995), Jean Langford (2002), Gyan Prakash (1999), and Ann Stoler (1995, 2002), among others have had much to say on the role and function of science and technology within the colonial civilizing mission, on "race," "caste," "ethnicity," and "tribe" as outcomes of colonial science, on the cultural norms and forms of reason at particular scales and temporalities, and on relations between metropole and colony across hemispheric divides. These writings have also firmly established the connections between science and the nation in projects of colonial and postcolonial rule.

Links between science and the apparatus of state (relations of knowledge and power, in other words) demonstrate their effects in the realm of identity, and science is a primary site where putative "incommensurability" is produced. Incommensurable identities, dependent upon the experience of subjection in colonial and neocolonial contexts, should be understood as the outcome of elite efforts (knowing and unwitting) to instantiate hemispheric divides, and of subaltern struggles against such hegemony. Stagist theories of history worked out during colonial expansion continue to have relevance for the way many think of the scientific and political legacy of the European Enlightenment in places such as Indo-

nesia, that are far away from its (supposedly pure) point of origin. Yet, I argue, those identities that depend upon their relation to reason and modernity (either EuroAmerican or elite Indonesian identities) should be seen as outcomes of the practice of science (among other things), not the points from which science commences.

Struggles over the idea of generativity and newness—origin, originality, origination—are important aspects of postcolonial difference. How do we confront the aforementioned accusation of scientific non-reality, or address the problem that science must replicate patterns established in Europe or America to be recognizably "real"? From where does "newness" or "reality" come, and who should decide? Accusations of repetition (or of non-reality) depend upon a simultaneous desire to instantiate elite identities while ensuring that difference is sustained (Fanon 1991[1952]). This is a no-win situation; Indonesian scientists in the context of the Togean project can be seen struggling with the structure of (though not the fact of) the mimesis and ambivalence that Homi Bhabha (1994) describes so well.

Although funding flows primarily from organizations in the North to the many sites of biodiversity conservation in the South, I argue that knowledges, rationalities, and natures in Southern biodiversity conservation cannot be understood through the language of assimilation or adaptation in the tropics of a project that originates in a more temperate climate. Rather, nature-making in the global South has been productive of the very form that has come to be known as biodiversity. For example, ICDP projects should be viewed as an idea originating in Southern peoples' resistance to wildlife conservation, rather than as the brainchild of Northern biologists for whom the ICDP is merely a formalized response. Along with many theorists of postcoloniality, I insist in this work that science is multidirectional in its rationales, causes, and effects; ideas flow back and forth across hemispheric, identitarian, and conceptual divides linking metropole with postcolony, center to margin.

What emerged from Indonesians' conservation biology and practice in the Togean Islands was a highly specified "nature," and a particular way of understanding "the human," that contributed not only to the transnational problematization of environmental risk known as biodiversity, but also to a domestic project of building the Indonesian nation-state. Many Indonesian scientists I encountered believed that the transnational discourse of biodiversity, with its particular notion that people who live near rare plants and animals threaten this nature, does not fit the circumstances of Indonesia very well. For example, at Camp Uemata, Budi stressed to me that conservation was a "Western" idea. He said, "A country like America is rich enough for conservation and looking at wildlife the way Americans do is a luxury. I don't think the Indonesian people are ready

to look at nature in this luxurious way. Nature is still full of resources for Indonesian people because of our level of development. We are not rich enough in Indonesia to afford conservation of species—species is a Western concept. Yet, conservation is an 'in' for me; my biology employs me."

I argue that a study of science neither "ethno" nor "Euro" entails a recuperation not only of the making of science's matter, but of where and by whom that matter is made. In the case of Indonesians' conservation biology, this requires an exploration of how Indonesian scientists were shaped by the Indonesian state during the period of authoritarian leadership of former President Suharto. It also necessitates study of the commitment possessed by many of these same scientists to look beyond the state to the affective nation. For instance, while Togean people were often viewed through the biodiversity problematic as those humans who threaten both nature and state order, at other times scientists recognized them as Indonesian citizens. In order to avoid being inhabited by the supposition that "there are no *real* scientists here," we will need to take seriously and examine closely Indonesian scientists' commitments to the nation form and their efforts to think through the meaning of their own, as well as Togean peoples', lives as Indonesian citizens and subjects.

A Thread of Rattan

The mountain at Benteng is the "navel" of the Togean Islands and through it the islands are connected to the wider world. Benteng is the site of an old fort built by Bugis and Sama people at their first site of habitation when they came to the islands from southeast Sulawesi sometime before the 1860s.[13] Puah Umar[14] related to me how in the old days the Togean Islands were called Togoya. Togoya was the youngest of three siblings. The eldest was Selatan (South Sulawesi), and the middle sibling was Ternate (an island and former sultanate to the east of Tomini Gulf). The three siblings were connected together by a giant rattan running through the earth: in Selatan were the roots of the vine; in Ternate were the main branches; and in Togoya, on Benteng mountain, you can still find the tips of the rattan emerging from the ground. Through this rattan vine, the Togean Islands were bound to the Raja of Bone in South Sulawesi, and to the Sultan of Ternate to the east.

Like any project of nature-making, the Togean Island biodiversity project would not be imposed on *terra nulis*. Nearly 30,000 Togean people of more than half a dozen ethnic groups (Sama [Bajau], Bugis, Saluan, Togean [Ta'a], Bare'e, Gorantalo, Bobongko, and others) lived in the islands in the mid-1990s (Kantor Statistik Kebupaten Poso 1995a and 1995b). All of the major Togean ethnic groups had come from the sur-

Benteng Mountain, by Celia Lowe.

rounding mainland, and Sama and Bugis people had migrated there as well from South Sulawesi via the eastern coast when Sama fishers arrived in search of turtle shell and sea cucumber and Bugis traders had followed them to buy their sea products. The economy of the Togean Islands still revolves to this day around exports of land and sea harvests. Most of these exports follow traditional subsistence and very old trade patterns, like the trade of sea cucumber to China, which has persisted for at least a millennium in eastern Indonesia (Warren 1981). But there is also a new Australian-Indonesian pearl farming venture in Kilat Bay, and a Japanese-Indonesian logging conglomerate has stripped the interiors of the larger islands of trees. Tourism is another industry that connects the Togean Islands to far-away places, in this case mainly to Europe, in a sustained and novel way.

The archipelago is divided into two administrative districts: Una Una and Walea Kepulauan. Every Togean village has an elementary school, and there is one middle school in each district. The majority of Togean villages in the 1990s were described by the government as "left behind" (*desa tertinggal*) and received government benefits based on this status. With few exceptions, the people of the islands follow Islam rather than any of the other four state-sanctioned religions and, despite the appearance of isolation, the Togean Islands have been connected to wider spiritual, economic, and political worlds for as long as there have been people living there. As the third line running across the archipelago, Puah Umar's thread of rattan symbolizes these connections: plants and animals have

always linked Togean people and their islands to centers of far-away charisma and influence.

Biodiversity is a new layer in this much older story. Rather than purchasing all kinds of plants and animals for sale and export, however, conservation biologists came to the islands to enumerate, classify, and preserve the creatures they found. While species inventories are imagined as the foundation of biodiversity conservation by Indonesian and international biologists alike, numbers and totals have proliferated in conservation projects around the globe with only a magical connection between inventory and any ability to preserve nature's newly visible creatures. There is no clear identity between crisis (loss of biodiversity) and form (enumeration). The wrench in the works, from a biodiversity perspective, is *human* incalculability. Biologists have been unable to save vast species inventories on their own; they require the help of a local populace, often the very same peoples whom they suspect of instigating biodiversity loss. Thus, biodiversity projects have focused their gaze on these populations and, alongside of Togean species, Togean people have become an object of scientific study and analysis.

My interest in Togean people shadowed the interests of the biologists, and I was most curious about people of Sama ethnicity for two reasons.[15] First, biologists from the Togean project were themselves very concerned about Sama people, often believing Sama are the ethnic group most destructive of the environment, especially the marine environment. Scientists frequently attribute the problems of blast and cyanide fishing to Sama people because Sama are renowned as fishers and many make their livelihoods from Togean waters and reefs. Since Sama people, wherever they are found (along the coastlines of Sabah, Malaysia, the southern Philippines, and throughout eastern Indonesia), have often followed sea cucumber and fishing harvests living aboard small boats, they have been called Southeast Asia's "sea nomads." This brings me to a more personal reason for my interest in Sama lifestyles. I myself had lived aboard a boat for seven years in the 1980s and had circumnavigated the earth passing through the tropical waters of Asia and the South Pacific. For this reason I felt especially intrigued to learn about the identity of an ethnic group described as "sea nomads" and "floating people" (*suku terapung*).

There are several Sama villages in the Togean Islands. These villages (*desa)* are usually found on outlying islets offshore from the larger of the Togean Islands and are readily identifiable by their bald white mounds stripped of trees where the soil has been used to fill in the surrounding shallows. Sama houses tend to be built on stilts, with the front edge standing over the water and the back of the house on dry or reclaimed land. Like other Togean Island people, Sama work at small-scale resource harvesting, gardening, coconut palm farming, and low-level trade, and their

activities are more varied than fishing alone. I developed most of my research relationships and friendships in Susunang, a village of around 2,000 people divided into five hamlets (*dusun*) and the largest Sama village in the Islands. I had a house built for me there, bought a canoe, and spent the better part of 1996 and 1997 traveling, talking, fishing, and farming with Puah Umar and many others.

Through these activities, I began to understand how Sama people had engaged with the materiality of the Togean Islands through their own historical and contemporary nature-making practices. As a counterpoint to the discourse and practice of scientific biodiversity, I followed in detail how Sama fishers and farmers physically and discursively constructed Togean land and marinescapes and made the islands into a productive and storied place of familiarity and knowledge, experience and expertise. As a result, I tell a different story about Sama people and their relationships with Togean flora and fauna from some of the conservation biologists I know. I do not want to claim, necessarily, that my descriptions of Sama peoples' lives are more "true" than those biologists tell. I do want to demonstrate, however, that we can tell a story of Sama peoples' nature-making that is at least as complex and nuanced as the narrative of how biologists have come to "discover" profuse Togean natures.

Part of this complexity entails how human mobility in the Togean Islands is narrated. While scientists travel the world frequenting conferences and seminars in far-away locales, attending to natures that are remote from the places where they themselves reside, and consuming natural resources from sources they can't always identify, most of the Sama people I met had never been outside of the Togean archipelago. Still, Sama travel aboard canoes on resource-collecting trips called *pongkat* is viewed by elite Indonesians as uncivilized, as is Sama peoples' habitation of stilt houses overhanging the water's edge. Sama "mobility" is perceived as threatening, both to biodiversity and to state control of the Togean population, and there have been several efforts to resettle Sama people from their offshore villages onto the land of the larger Togean Islands. At the same time, paradoxically, Sama and other Togean peoples' perceived lack of cosmopolitanism and connection to a wider world is also seen as a threat to biodiversity and governmentality. This is the case even though they are obviously integrated into larger scales of region, nation, and cosmos in complex ways.

Togean people, Sama people in particular, are crucial then to the story of conservation in Indonesia. If biodiversity is a new problematization of the relationship between nature and the human, we will want to know who is this "human" that appears to trouble nature so. The figure that Indonesian scientists produced through their Togean Island work was inevitably a different human than that developed by EuroAmerican biolo-

gists in thinking through the problem of tropical biodiversity. This is because Togean Island people were necessarily more than just "threatening humanity" to Indonesian biologists. They were also Indonesian subjects and citizens who fit into recognizably Indonesian ideas of national development, belonging, and alienation. The particularity of Indonesians' biodiversity was dependent on this understanding.

For this reason, I have found it useful to follow Sama people using the same methods I use to track scientists in their work. I ask, What does Sama peoples' nature-making entail in the Togean Islands? What are the networks that connect Sama people across scales of region, nation, and transnation? How far do Sama peoples' inscriptions, assertions, and objects travel? And how are Sama logics similar to or different from authorized and "official" discourses of science, nature, or nation? The story of mid-1990s Togean biodiversity cannot be comprehended without this parallel narrative of the ways Togean people engaged with the biodiversity project and with their own material and idealist natures.

The Reason for Reason

Indonesians' encounters with the biophysical world have conventionally been described and evaluated from the perspective of "indigenous knowledge." At times this has proved a hopeful and necessary strategy, and the framework of indigenous knowledge has been crucial to expanding our understanding of what is able to count as knowledge. Different, so-called "indigenous," ways of knowing the environment have often proved commensurable in relation to the hegemonic knowledges of natural science, and are frequently demonstrated to be more efficacious when compared with the nonsituated and instrumental knowledges produced and promoted by governments or transnational modernization schemes. Indigenous knowledges have also been used to reflect back upon natural science. In the process it has become clear that all knowledges are situated, practical, dispositional, flexible, and social.

I have chosen to study Indonesian encounters with nature not through the analytic of indigenous knowledge, but rather through an analytic of "reason." Obviously, even though Indonesian scientists are literally as "indigenous" as Togean peoples, they do not resemble the figure who forms either the conventional subject or object of indigenous knowledge; Indonesian scientists have represented Togean nature through "transnational," not "traditional," ways of knowing. I also find it more problematic than productive, at this point, to describe Sama peoples' ways of knowing nature through the rubric of indigenous knowledge. If all knowledges are situated, practical, dispositional, flexible, and social, then we

need an analysis capable of placing Indonesian scientists and Sama people in the same frame. While scientists' knowledges are clearly not indigenous, Sama knowledges have cosmopolitan dimensions that also do not fit within the ambit of indigenaeity.

Likewise, there are problems with a conventional tracking of "knowledge." "Local" knowledge has often seemed an antidote to the universalist claims of scientifically derived natures, or the global ambitions of biodiversity conservation. The assumption in this work has been, if you know a group's boundaries you can investigate their knowledge, coherent and original. Natural scientists themselves are also presumed to be a group, bounded and whole, who share a single form of knowledge and a common worldview. As such, the study of knowledge has often been about the collection of objects—bits and pieces of information or perspective—rather than an examination of knowledge-making projects or the travels and transformations of thought across space and time. The science of ethnobiology, to cite but one example, has often entailed collecting names for things and interpreting these lexicons as coherent knowledge. Such a collection does not capture nonlexical hybridity, change, or process, however, or the elements of thinking as practice.

While "knowledge" proposes fully formed objects that are simply excavated or revealed, "reason," on the other hand, delineates the active practice of thinking. In Paul Rabinow's (2003) words, thinking "is a situated practice of active inquiry whose role and goal is to initiate a movement from a discordant situation to a more harmonious one." Knowledge might change through learning or discovery—the acquisition of more objects—yet, thinking demands the assemblage of matter, language, and technique—the creation of new objects. Since the time of Aristotle, studies of reason have interrogated the status of objects and how claims to knowing are made. I engage this process here, not to determine the foundations of truth sought by ancient Greek or Enlightenment thinkers, but rather to understand the conditions of possibility upon which particular contemporary thinking and knowledge-making practices rest.

Although a study of knowledge can answer many of our questions about nature, it does not help us to see how nature becomes a question in the first place. A study of reason, on the other hand, operates as a metacommentary on knowledge. It shows us what will become valued as knowledge, how thought is actively structured and normed, and how these norms expand to cover a variety of situations. More than a hemispherically bounded analytic, reason can help to uncover the possibilities for, and conditions of, thought no matter where thought is located. This is not how reason is ordinarily conceived of by those who assume for themselves the mantle of rationality, however. Reason is conventionally understood as a system of universal truth that reveals the natural traits,

and the relations between the elements of nature, that allow us to apprehend an actually existing world "out there." Kant (2001[1781]) famously critiqued this concept of "pure" reason for the proposition that the truth of an object in and of itself is discernable. Rather, he argued things can become objects of knowledge only with recourse to the experiencing human mind structured a priori by space and time.

An archaeological shift in the milieu of reason occurred at the start of the nineteenth century, when the European colonial enterprise, formerly organized exclusively around extractive trade practices (those, for example, of the Dutch and British East India Companies), became reorganized around the governance, education, and "advancement" of colonial populations. In the process, the capacity to recognize and acknowledge the universally true and good came to be viewed as unequally distributed along social lines of race, class, gender, and geography. Within both metropole and colony, reason was claimed as the property of a certain class of white European men, while women, peasants, laborers, and non-Europeans embodied backward, unchanging tradition as well as enchantment, emotion, and sentimentality. The civilizing mission (the success of which was, of course, infinitely deferrable) emerged as a technological remedy for the lack of reasoning capacity among the white man's Other.

Alongside these historical developments, the origin of reason in nature and its teleological progress in human history began to be called into question in European political philosophy. Nietzsche (1998[1887]), for example, described how reason not only reveals things but shapes them, and he commented on the reality-shaping effect of affect: "[T]here is *only* a perspectival seeing, *only* a perspectival knowing; and *the more* affects we allow to speak about a matter, *the more* eyes, different eyes, we know how to bring to bear on one and the same matter, that much more complete will our 'concept' of this matter, our 'objectivity' be" (85). Weber (1946[1914]) wrote of bureaucratic reason and its capacity to bound up human creativity in routinized and stultifying forms that disenchant the world we live in. Writing as refugees from European Fascism, Horkheimer and Adorno (1969[1944]) saw the failures of reason as part and parcel of the failures of capitalism, and argued that as Enlightenment attempts to secure itself against the mythic it becomes as totalitarian as any system. And Freud (1989[1933]), while espousing the scientific *Weltanschauung*, was instrumental in exposing the human mind as divided against itself, thereby opening up new possibilities for a theory of subjectivity.

Most nineteenth- and early-twentieth-century thinkers maintained liberal positions on reason and its capacities—if only reason could be rid of one tainting imperfection or another it might still perform its liberating function. It is also worth noting that nearly all placed reason in Europe using Asia, especially China, as a site of originary backwardness against

which European reason might be recognized. By the late twentieth century, however, reason had come to be seen for its arbitrariness—as a repository of social norms and ancient prejudices, rather than as a guide to universal, unsituated truth. As elaborated in the work of Canguilhem, Foucault, Rose, or Bourdieu, for example, reason displays a self-interestedness that situates it within a field of normalization and social power. Instrumental, calculating reason, including, especially, "reason of state," employs the proposition of universal validity in order to reinforce the claims of many narrower interests. Critical reason, on the other hand, is a way forward, if not a way out.

The status of the object has also been in flux. How do we know the object if it is conceived differently from different subject positions? Following upon Marx's insistence on the objective value of proletarian vision from below, theories of situated and partial knowledges have emerged (Hartsock 1999). Pressures on the postcolonial thinker, women, and subaltern subjects to conform to the God's-eye view of universal knowledge as seen from patriarchal centers have been elaborated (Harding 1998). At the same time, the argument for situated knowledge has also been an attempt to get closer to the object, not to do away with the principle of objectivity entirely (Haraway 1991). Anna Tsing has made an especially compelling argument for "engaged universals" observing that universals can be found in both imperial and liberatory projects: "Universals are effective within particular historical conjunctures that give them content and force. We might specify this conjunctural feature of universals in practice by speaking of engagement. Engaged universals travel across difference and are charged and changed by their travels. Through friction, universals become practically effective. Yet they can never fulfill their promise of universality. Even in transcending localities, they do not take over the world" (Tsing 2004:8).

Although universal human reason has come under revision in political theory and critical ethnography, this reason still maintains an active life as a folk category. While anthropologists and others now tend to view reason as a *strategy* of universalization, in certain places, among certain types of institutions, among certain peoples at certain times, to deny the universal means to deny one's own humanity (Bourdieu 1998:89–90). Natural science is the social field perhaps most invested in the universality of reason, where to disclaim a truth derived through scientific practice is to become a lesser form of human. Nonetheless all concepts, scientific or otherwise, have particular historical and social conditions of possibility. Dipesh Chakrabarty (2000) has argued for "provincializing" universalist claims in EuroAmerican thought by examining these enabling conditions. Reason, as a folk category, entails those ideas, forms, and assemblages

that attempt to erase their own conditions of possibility to stake universal claims, yet histories of thought as practice, and the locations from which claims on the universal are made, leave traces. *Wild Profusion* is a study of these traces.

To understand how biodiversity was a new assemblage in Indonesia, how it problematized relations between Togean natures and Togean peoples, and how at the same time scientists themselves operated within degrees of freedom and constraint, it is helpful to understand biodiversity as a form of reason that produces both subjects and objects. Thinking of humans as subjects constituted by knowledge, rather than unconstrained individuals who possess and control knowledge, is useful for comprehending the identities that emerge from projects to conserve biodiversity. For example, at some point in the 1980s Indonesian activists (and here I include scientists with ethical and interventionist mandates) changed the designation of their collectivities from "nongovernmental organizations" (*organisasi non-pemerintah*) to "peoples' self-improvement leagues" (*lembaga swadaya masyarakat*, LSM). We see here how elite Indonesians were invited in the Suharto era to become a prosthesis of the state, not a node of opposition to it. This act involved an arbitrary reorganization of the self designed to accommodate the state's anxiety toward anything exterior to it. While Indonesian scientists' mediation between the state and Togean people was regulated and normalized, scientists themselves were also in the business of producing normalizing and standardizing knowledges about Togean people. These knowledges might serve state interests, though just as easily they could become a way to challenge the reason of state.

In the move from indigenous knowledge to an analytic of reason, I seek, therefore, to describe the contingent conjuncture where science, nature, and the Indonesian nation came together through a particular problematization of nature and the human assembled under the sign of biodiversity. The three lines running across this text—Wallace's line representing Sulawesi's unique faunal forms, the equator representing hemispheric divides across the fabric of human identities, and a line of Sulawesi rattan symbolizing Togean peoples' connections to cosmopolitan worlds through plants and animals—intersect to create a circumstance within which transnational biodiversity conservation gained specificity, form, and generative capacity.

In reflecting upon the possibilities and limits of such forms and capacities through genealogies of reason, I hope to introduce new possibilities for thought, moving from discord to harmony, therein opening up a space from which to understand the lives of Indonesian scientists, Togean people, and the profusion of Sulawesi flora and fauna.

The Lines of This Book

The narrative of *Wild Profusion* is divided into three parts. Part 1, "Diversity as Milieu," moves from the forms of transnational biodiversity to the specificities of Togean nature/cultures to looks at how "biodiverse nature" and Togean "indigenous knowledge" were outcomes of conservation biology. Part 2, "Togean Cosmopolitics," proceeds in the opposite direction. Beginning from Togean natures, it examines how Sama people produced local land and marinescapes as well as cosmopolitan worlds through their nature-making practices. Part 3, "Integrating Conservation and Development," studies the outcome when the world of biodiversity science and management meets Sama people's nature-making.

In chapter 1, "Making the Monkey," I explore the postcolonial world of Indonesian science by examining how the Togean macaque (*Macaca togeanus*) was proposed as a new species endemic to the Togean Islands. Through the scientific practice of conservation biology, Indonesian primatologists identified the monkey first as a "new form," then as a "dubious name," and ultimately as an "endemic species." Throughout these acts of making, unmaking, and remaking the monkey, its unique status was crucial for developing Indonesian science, attracting foreign donors, and enlisting government and public interest in Togean Island nature, even as the natures important to Togean peoples were overwritten in the process.

Chapter 2, "The Social Turn," asks how Indonesian biologists formed their object of reason when this object was no longer conceived of as "nature," but rather as "culture." I begin with a comparative perspective on past and present instances of nature-making in the Togean Islands and eastern Indonesia. Adolf Bernard Meyer, who collected Togean birds in the 1880s, and Georgius Everhardus Rumphius, who collected natural curiosities in seventeenth-century Ambon, each developed a particular representation of Malay people that differs from the figure of the human produced in biodiversity conservation. Through a "social turn" in 1990s conservation, Togean people were asked to contribute "indigenous knowledge" to the project of conserving Togean nature. "Ethnobiology" and "participatory spatial planning" became the mediums through which Togean peoples' knowledges were elicited, and through which their identity as "indigenous" emerged.

I explore how it is possible to think Sama identity and Togean nature simultaneously without essentializing such a connection in chapter 3, "Extra-terrestrial Others." Sama identities and natures have taken on cosmopolitan attributes through their connections to, and imaginations of, other places. I follow these connections through Sama place-making, including a sea cucumber collecting trip, making sago in a sago palm

swamp, through place names, and by way of canoe travel along the Togean shoreline. I then contrast these practices with the "tribunals of reason" Togean Sama people are subject to and that frame Sama lifestyles in negative terms.

In chapter 4, "On the (Bio)logics of Species and Bodies," I look at how Sama identity was produced in relation to enchantment, disenchantment, and instrumental reason in the context of health and well-being in the Togean Islands. Both biomedicine and biodiversity conservation have been similarly viewed as remediation for "improper" or "ineffectual" knowledge or belief, for pseudoscientific practices, and for social "underdevelopment." Managerial interventions in the Togean Islands often took on unexpected forms, developed unanticipated lives of their own, and ultimately missed their mark. In this chapter, I extend my analysis of nature to questions of bodies and health as they intersect with Togean land and marinescapes for Puah Lidja, her family, and her neighbors in Susunang village. Further, I describe how new forms of "unreason" reveal the limitations of and aporia in practices of scientific calculation and instrumental governmentality.

In chapter 5, "Fishing with Cyanide," I present one specific type of environmental threat, cyanide fishing practiced in the live reef food fish industry, and interrogate it for what it can tell us about identity, science, and legal rationality. Most biologists understood Sama fishers to be the perpetrators of cyanide fishing, and Indonesian laws and conservation philosophies rhetorically placed fishers at the center of responsibility for how fish were caught. Many others of diverse ethnic backgrounds were also involved in the live fish trade however and, in most cases, at a more fundamental level. Chapter 5 examines how "community" became the arena for conservation and bureaucratic intervention despite the cosmopolitan nature of the live fish trade.

Chapter 6, "The Sleep of Reason" documents the ultimate emergence of the Togean Island National Park in 2004. Serving as an epilogue to *Wild Profusion*, this final chapter also describes change in the form of 1990s biodiversity conservation—which coalesced around species, the ICDP, the emergence of Southern expertise, and late-Suharto era political norms and cultural forms—to a new millennium milieu emphasizing ecoregions, enforcement, and the possibilities of the post-Suharto "reformation" era. Reason, I conclude, produces monsters when it is purified from enchantment. Reading conservation practice through Francisco Goya's 1799 painting, *El Sueño de la Razon Produce Monstruos*, the rational mind and its demons meet, amidst the seemingly pure logic of disappearing species and the imaginations that are necessary to preserve a diversity of life as well as lives worth living.

PART ONE

Diversity as Milieu

Akuna Pongkat *Dan pergilah masyarakat Bajau ke laut, jauh. . . . Kehadiran ikan paus merupakan tertanda datangnya musim ikan.*

I Pongkat And go ahead Bajau people to the sea, far away. . . . The presence of whales is a sign that the season of fish has come.

Kuda Laut *Eksotisitas Indonesia di mata dunia sebagian terpenting adalah pada lautnya.*

Sea Horse The exoticness of Indonesia in the eyes of the world for the most part is related to the sea.

Paka Lele dan Sawi *Kehidupan mereka masih di warnai oleh corak tradisional. Mereka telah mengikatkan diri pada ikatan sosial yang menonjol pada tindakan kolektif dalam satu komunitas.*

Paka Lele and Sawi Their lifestyle is still colored by a traditional patterns. They are already connected by social ties conspicuous for collective action in the community.

Ilmu Bajau *Sebuah kampung Bajau terhambur di atas barisan karang. Di tengah birunya kepulauan Kaledupa. Kampung ini adalah tempat terakhir untuk menikuti kehidupan dan 'misteri' orang Bajau.*

Bajau Knowledge A Bajau village scattered atop a row of coral. In the middle of the blue, the Kaledupa Islands. This is the last place for finding the lifestyle and 'mystery' of the Bajau people.

Koin Etnik *Hasil laut sejak dulu jadi komoditas orang Bajau. Hasil laut itu kemudian mereka tukarkan dengan 'nilai' yang telah disepakati oleh kedua pihak.*

Ethnic Coin Sea products since early times have been a commodity for Bajau people. They trade these products with a 'value' that is already agreed upon by the other party.

Rajah Bajau *Ungkapkan bahasa rupa dari reka hias Bajau. Sebuah simbol rupa runggu tradisional.*

Bajau King Speaking their language is one form of Bajau creativity. A symbol of their tradition.

—Sopandi, Jelajah Etnik

In retrospect, Indonesians were rethinking diversity in relation to both nature and nation during the waning Suharto years. On the one hand, national norms for nature and its uses were being called into question. Were Indonesia's trees and minerals to be a resource for logging and mining and other forms of elite national development, or was Indonesian nature a resource for "the people" (*rakyat*) to create a healthy subsistence? Parameters of social inclusion and exclusion in the nation were likewise under revision. Would acceptable cultural difference continue to be narrowly defined by the modernist state, or might new forms of identity be folded into the nation's understanding of itself? This national conversation on diversity is the milieu within which scientists' species inventories and their study of indigenous knowledge in the Togean Islands can be understood.

The working through of the problems of ethnic and natural diversity can be seen in two different gallery exhibitions for which Sama people (who are called "Bajau" in these works) provided the raw material. Sama were fit simultaneously into the twin configurations of ethnic teleology and nationalist history in the 1990s. First, they were considered an "alien ethnic group" (*suku terasing*) at a moment when hemispheric divides were constructed between Indonesian margins and centers at the intersection of ethnicity and a particular New Order framing of acceptable cultural difference. Second, Sama were considered a "marine ethnicity" (*suku laut*) at a time when Indonesia was beginning to reconsider its maritime heritage. These contingencies explain the "discovery" and "display" of Sama people as a resource for a national conversation on nature and identity.

The first of the exhibitions, *Bajau*, was scientific, ethnographic, and educational in nature, and was an outcome of a scholarly conference on Bajau/Sama communities held at the Indonesian Institute of Science (LIPI) in Jakarta in November 1993. The three-day conference, "addressed the re-introduction of Indonesia's cultural and ecological diversity as national assets" (Sejati 1994:34). The scientists in attendance all were scholars of Sama peoples' "ecological adaptation, nautical skills, resource management, maritime wisdom, and particular sea lore." The exhibition itself was constructed around a replica of a Sama village set in a water reservoir and filled with floating canoes brought from Sulawesi. Around this centerpiece were placed displays explaining the distribution of Sama communities around Southeast Asia; Sama origin stories and tales of life on the sea; terminologies and lexicons in Sama language; a description of the Sama environment; a story called "A Day in the Life of the Bajau"; a

description of sea cucumber collecting; and an explanation of Sama medical practice and belief.

I was not in Indonesia in 1993 for *Bajau* and know it only through conversations with its curators and through its catalogue. In June of 1997 I was fortunate to witness a second exhibition however, in which Sama ethnic identity was invoked as a national resource. This exhibition, *Jelajah Etnik [Ethnic Explorations]: A Journey Through Wallacea*, held in the lobby of the Jakarta World Trade Center, presented a series of paintings by the artist Sopandi on the theme of Sama life in the Wallacea region. Sopandi's paintings were a bricolage of hornbills, wild orchids, buffalo, dragons, tuna, boats, sea shells, and spirits, set in fields of abstract shapes. Each painting was overlain with intricate pen and ink line drawings and filled in with watercolors. Bursts of *mega mendung* cloud patterns, inspired by Javanese batik, formed the backdrop for the wild activity in the foreground. The finished canvases were framed with carved hardwoods described by the artist as "ordinary firewood."

The *Jelajah Etnik* exhibition catalogue presented the artist in baggy khaki hiking pants, T-shirt, and canvas hat carrying a large backpack and bedroll. Looking like a nineteenth-century explorer, his image is superimposed upon a reproduction of Alfred Russel Wallace's 1868 map of the Netherlands East Indies. Treking across the map below Sopandi are eight nearly naked Papuan people carrying machetes and wicker backpacks, seemingly on their way home from tending a garden. The catalogue is filled with descriptions of the paintings and of the artist's adventures traveling in Southeast Sulawesi.

Whether as science or art, the exhibitionary imagination is always political.[1] The "native village" display, like that in *Bajau*, is a trope of ethnic exhibition deriving from late-nineteenth and early-twentieth-century universal expositions and world fairs. By presenting "native villages" and "native peoples" as *tableax vivant* spectacles, ethnic exhibitions educated EuroAmericans in racial and cultural superiority (Barkan and Bush 1995:25). Designed with the best evolutionary science of the time, one famous example, the Philippine exposition at the 1904 World's Fair in St. Louis, aimed to re-enforce U.S. imperialism after the Spanish-American War. This display brought to life the notion that the Philippine people were civilizationally inferior and incapable of governing themselves without help from the more "advanced" United States.

Bajau, likewise, was a spectacle of elite Indonesian progress and superiority, and the evolutionary preoccupation familiar from the universal exposition is reflected in this description from the exhibition: "Today, the seafaring culture of the Bajau remains an example of these early maritime communities. Indeed their present day practices are direct links to Indonesia's maritime past" (Sejati 1994:3). Similarly, the museum's "archaeolo-

gists," those in charge of Indonesian prehistory, oversaw the exhibition's installation. Indonesian and international scholars alike pursued "ecological adaptation" as the modality for describing Sama peoples' lives, a conceptual approach that fit well with the internal colonialism of the Suharto state.[2] In *Bajau*, Sama people are a living anachronism.

Like the paintings of Picasso, Sopandi's style belongs to the twentieth-century tradition known as "primitivism," and represents a romantic encounter with the exotic, unfamiliar, and anachronistic. The figures who walk across the pages of the *Jelajah Etnik* catalogue are racially Papuan, not Malay like Sama people, reflecting an Indonesian racial formation that associates darker Papuan features with primitiveness. Similarly, the Javanese batik Sopandi employs in the background of his paintings are iconic representations of ethnic difference in Indonesia. Sopandi presents his expedition to Wallacea as cultural time travel and his imagination is haunted by ninetenth-century colonial exploration.

Here is the question though. Must we read both of these exhibitions merely in the context of turn-of-the-twentieth-century European evolutionary thinking, or can we also understand these Indonesian scientists and artists as attempting to remedy problems other than those solved by earlier native village displays and primitivist art? What if we were to read *Bajau* and *Jelajah Etnik* against the grain, as thought that searches for new solutions to the problem of Indonesian modernity, albeit steeped in familiar modes of representation? In such a symptomatic reading, the place of ethnic and natural diversity within the nation appears, not merely as determined ideology, but as a problem to solve. While the two exhibitions each propose Sama inclusion into the nation on elite terms, more radical possibilities for the meaning of inclusion work to subvert the obvious exclusionary readings.

For example, programmatic activities during *Bajau* describe the fully modern political and environmental problems experienced by Sama people as citizens of the national polity. In a section of the exhibition catalogue titled "Issues Affecting the Bajau Today" the curators wrote:

> In collaboration with local Bajau researchers, LIPI and Sejati research presents the ecological and social-cultural issues affecting the Bajau today. Contributors such as the Asia Wetlands Bureau, UNESCO, and other institutions also provide books and articles on marine resources. In this room, the visitor can learn about how changes in marine ecology influence the whole of Bajau society. The visitor could study how marine (traditional) resource management would be applied to wider development projects. Equally, the collected research would show the commercial potential of marine resources, and this includes marine tourism, for Indonesia's national development. Most im-

> portantly, the main hosts of the exhibition, two Bajau representatives from North Sulawesi, would always be present to answer questions from the public and to recount their immediate experience. (Sejati 1994:38)

Public programming for *Bajau* included a discussion with Abdurrahman Wahid (leader of the Muslim organization Nahdlatul Ulama, who would later become Indonesia's fourth president) on the importance of cultural diversity in Indonesia; a visit by fishermen from Jakarta Bay to discuss issues they shared in common with Sama representatives; and special events for business leaders and school children. Despite its anachronistic representational form, *Bajau* contains a sub- or parallel text that challenges the thinking of Indonesian elites, "creating something new within the most traditional political forms" (Rose 1999:280).

Similarly, while Sopandi's style belongs to the early twentieth-century tradition in painting and sculpture called "primitivism," the context for his work is not early twentieth-century Europe. In his paintings and catalogue descriptions, Sopandi presents a picture of Sama life that is very different from the Indonesian state's own evolutionary representations of Indonesian life outside of Java, Sumatra, and Bali. For example, he describes his painting *Bajau King* with the caption, "Speaking their language is one form of Bajau creativity. A symbol of their tradition." Clearly encompassed by a German Romantic theory of language, the caption nevertheless might also be interpreted as contesting state language policy that promotes the national language, Bahasa Indonesia, over regional ethnic languages. The value Sopandi places on Sama language confronts the state's rationalist desire for Indonesians to speak formal "good and correct" Indonesian (*bahasa Indonesia baik dan benar*). In calling Sama language a form of "creativity," Sopandi cracks open the universalist logics of governmental reason.

In order to understand how Indonesian scientists produced Togean Island nature and culture, we need to understand something about the milieu of diversity they were working within at the time. In this context, I propose we will learn more about Indonesians' scientists and their work by taking them seriously as honest brokers struggling with what diversity can and will mean in the context of late-twentieth-century Indonesia. Although we cannot fail to recognize the power elite Indonesians have to represent, and while we can read both rational evolutionism and imaginative Romanticism into these exhibitions, *Bajau* and *Jelajah Etnik* were attempting to solve other problems than were turn-of-the-twentieth-century European expositions or exhibitions. Barkan and Bush, in their exploration of primitivism as a particular form of modernism, claim: "As

primitivism reappeared in text after text, each new ideological mix proved unpredictable" (1995:13).

In part 1, IFABS scientists can be seen grappling with questions of how to represent Togean biodiversity and Sama identity. Questions about Togean species emerged from within transnational conversations on conservation biology and national discussions on science and nation. Questions about Sama peoples' "indigenous knowledge" arose from within the milieu of *Bajau* and *Jelajah Etnik* where the meaning of diversity within the nation was at stake. These active practices of thinking produced such objects as "nature" or "culture" that should be understood outcomes, not starting points, for Indonesians' science in the Togean Islands. From within this milieu we can hear scientists ask: What will count as the value of natural diversity for the Indonesian people?; What will constitute acceptable norms of cultural difference within the nation?; and, How can these values and norms be represented?

Chapter One

MAKING THE MONKEY

> The [Togean] animal kingdom is, as cannot be expected otherwise, poor in representatives. It is said that the only mammals living here are bats, rats, and the *babi rusa* [deer pig]. Of birds, we find many of the species living along the main coast [of Celebes]. On our walk through the main village I saw *Trichoglossus ornatus*, *Tanygnathus megalorhynchus*, and *Nectarinia lepida*. There are few snakes and few crocodiles, and turtles are only found near Poeloe Sendiri. The sea between and around the islands is also poor in fish, a phenomena certainly worth mentioning. On the other hand, the sea surrounding the islands is rich in holothurians [sea cucumber], the most important article of trade and export in these islands. Finally, we noticed on our walk the most beautiful land snail, a Nadina, which we had not yet seen on the main coast.
>
> —C.B.H. von Rosenburg, *Travels in the Region of Gorontalo*

TOGEAN ISLAND biodiversity was not at all self-evident in the beginning of the 1990s. Nor was the archipelago's appropriateness as a new national park. In order for the Togean landscape to move from "poor in representatives" (as it was in 1865) to "rich in biodiversity" (which, by the mid-1990s, it had become), the "facts" of Togean biodiversity awaited their empirical demonstration and social emergence (Latour and Woolgar 1986; Shapin and Schaffer 1985). Such a representation of Togean nature was encompassed by the work of species inventory in the emergent field of conservation biology. Key to the appearance of biodiverse nature in the Togean Island was the Togean macaque, *Macaca togeanus*, a primate living in the upland forest of Malenge Island. Nonhuman primates have always held a fascination for biologists due to their role in the history of human evolution. The reason of the moment proposed that if biologists were able to stabilize the species status of *M. togeanus*—if they could prove it to be unique and endemic—a protected-area initiative would be justified.

My familiarity with the Togean monkey and the question of its species status developed through my friendship with Jatna Supriatna. Dr. Supriatna, a conservation biologist from the University of Indonesia (UI), is the world's leading expert on the evolutionary biology and systematics of

Sulawesi macaques. In the mid-1990s, *M. togeanus* became a focal point for both Supriatna's research, and for the establishment of a more encompassing conservation program in the Togean Islands. As Dr. Supriatna observed in his discussions with me on species diversity, the species concept (rather than ideas of ecosystem, ecoregion, or environmental justice, to name just a few other possibilities) was crucial, both in biological and social terms, for saving nature. "Species are the key," he argued, "But there is a flexible concept of species. For example, think of the Javan rhino. Without species conservation, maybe Ujung Kulon [a park in Java] would never be visited by people, there would be no donor that likes to give money, there would be no government attention. Charismatic animals allow people to want to save the environment. People don't just say 'I want to save biodiversity.' Species are attractive."

In order to put together a Togean conservation project, Dr. Supriatna would use species to attract an Indonesian and international public, a foreign donor, and government agents and agencies. The existence of a unique Togean flora and fauna would entice domestic and international tourists, would create political support for a park among regional and national bureaucrats, and it was also necessary to secure investment from Conservation International (CI). In the process of making the monkey, "*M. togeanus*" would evolve from "new form" to "dubious name," and then reverse its course to become the "endemic species" the project required.

In cultural studies of science and technology we always insist that inquiries into the ways science is thoroughly imbricated with, and productive of, both nature and society is not about "good" and "bad" science. As a science studies scholar, I argue that Indonesians' science, like any other science, can show us how those bio-objects we want to call "natural" are immanently social and cultural as well as biological and physical. I am interested in the many positivities of the Togean macaque research, including the ability of a new and endemic species to form social relations and subjectivities within biological research and conservation. While the making, unmaking, and remaking of *M. togeanus* helps us to understand "sameness" and "difference" in Indonesians' biological science, my analysis should not be read as any kind of a commentary on the "quality" of the science as such. The monkey is, rather, an entry point for examining the intersection of science and nation within biodiversity conservation.

A New Form

The first hint of a distinct Togean primate species appeared in 1949 in an article by H.J.V. Sody, "Notes on Some Primates, Carnivora, and the

Babirusa from the Indo-Malayan and Indo-Australian Regions." Sody, a naturalist from Amsterdam who was trapped in Java by the Japanese occupation in World War II, used his time to make studies of the Bogor Museum's natural history collections, including Sulawesi macaques. He had available to him 17 skins, 15 skulls, and 1 stuffed specimen of unknown provenance. Sody combined an analysis of these specimens with the literature on 40 other skulls including a Malenge Island series collected by J. J. Menden in the Togean Islands in 1939. In October 1949, not long after Indonesian independence, Sody published his synthesis in the journal of the Royal Botanic Gardens at Buitenzorg (Bogor). He found that the skull index for macaque males was larger than for females and was the greatest in Menden's Malenge series. On the basis of its larger size, Sody proposed a "new form," which he named *Cynopithecus togeanus* (Sody 1949).

The macaque next appeared in 1969 in *Taxonomy and Evolution of the Monkeys of Celebes*, an account of Sulawesi primates by Jack Fooden. Fooden acknowledged that the Malenge monkey was morphologically larger and paler than the Tonkean (not Togean) macaque (*Macaca tonkeana*). The data did not convince him, however, that the Togean primate was a unique species. Rather, Fooden believed that the difference between the macaques of Malenge and the mainland population was probably caused by "relatively recent evolution in isolation," and he found "no characters that warrant allocation of the insular form to a separate species, or even sub-species" (Fooden 1969:114).

In the divergence between Fooden and Sody's opinions the species status of the Togean macaque became amenable to further study and empirical verification. In 1988 Dr. Supriatna was conducting a field study of hybridization between the Moor macaque (*M. maurus*) and the Tonkean macaque using behavioral and morphological analyses, for his doctoral degree in biological anthropology at the University of New Mexico. During his research travels with the "Sulawesi Primate Project," a name he had given to a team of Indonesian and EuroAmerican primatologists including several of his own biology students, he heard a rumor of a macaque population living in the Togean Islands. This inspired his first trip to the archipelago and resulted in an encounter with the Togean monkeys.

Coincident with the opportunity to define a unique macaque species, the emerging paradigm of biodiversity was beginning to open up new programmatic opportunities for conservation around the species concept. The Togean Islands had been identified as a potential conservation area by the UNDP Food and Agricultural Organization (Salm et al. 1982), the Asian Development Bank (1992), USAID (Soekarno 1989), and the World Wildlife Fund (Djohani 1989), among others. At the same time, Supriatna had developed ties to Conservation International through connections he

had made in the world of primate studies. In order to interest CI in a Togean Island project, his interests and his science would need to align with those of his funder.

Dr. Supriatna first began to sponsor his students to work in the Togean Islands on the macaque question in 1992 and 1993. During this period, Budi, along with several other students from the University of Indonesia, went to Malenge Island to study the macaques' behavioral patterns as well as Togean forest ecology. With Supriatna's backing, Firman, an undergraduate student from As-Syafi'iyah Islamic University, surveyed the size and constitution of the Togean macaque population. Firman not only found fourteen groups of macaques and some lone males, but he also determined that the monkeys were eating a mixture of fruits (mostly figs), leaves, flowers, grains, and seeds (Firman 1994). This early research led to the formation of the Indonesian Foundation for the Advancement of Biological Sciences (IFABS), an independent biologically focused NGO. Supriatna would be able to use the Togean project, not only for Indonesian nature conservation, but to develop an internationally recognized program of biological research. IFABS built its research station, Camp Uemata, to facilitate the primate research and to further their national scientific agenda (Surjadi and Supriatna 1998).

International interest in the Togean macaque research developed at the fifteenth International Primatological Society meetings held in Bali in August 1994. At the conference, the project to "clarify the taxonomic position of the Togean Island macaque" became a "priority action" (Bynum 1994). Immediately after the conference ended, Russell Mittermeier, president of CI and a primatologist himself, visited the Togean archipelago with Supriatna and other IFABS staff. This visit brought Indonesian scientists and their project together with a funder under the banner of preserving what CI began to refer to as the "endemic Togean macaque" (Mackie 1994).[1] CI then granted preliminary funding to support the new IFABS organization and the Uemata field station.[2]

While the IFABS biologists continued to study the Togean macaque in the field, Supriatna worked with his former advisor, Jeffrey Froehlich, on analyzing museum specimens. In 1996 they published their first scientific article on the species status of *M. togeanus* (Froehlich and Supriatna 1996). Employing dermatoglyphics, a technique of determining relatedness through fingerprints pioneered by Froehlich, and by examining the bathymetric contours of the Gulf of Tomini to determine that the conditions necessary to produce a species isolate could have existed, they elaborated three possibilities for the primate's taxonomic status: (1) *M. togeanus* is a third subspecies of *M. maurus*, (2) the Togean monkey is not subspecifically distinct from *M. tonkeana* (Fooden's hypothesis), or (3) the primate deserves separate species status (Sody's hypothesis).

On the strength of characteristics distinguishing *M. togeanus* from *M. maurus*—a larger cranium and diminished body size, longer tail, greater sexual dimorphism, unique pelage (coat), and varying dermatoglyphic patterning—the authors made an incipient case for separate species status. They argued, "comparably diagnosable pelage patterns, distinct alterations in body shape, and greater dermatoglyphic differences suggest that *M. togeanus* may be a valid species" (Froehlich and Supriatna 1996:65). Emphasizing the interpretation that supported separate species status, they proposed further research on the primate and an urgent conservation agenda. The "nonexistent protection and the tenuous status of the only known population of *M. togeanus* on Malenge Island lends urgency to the confirmation of these predictions and the formulation of adequate conservation initiatives" (66). As the facticity of *M. togeanus* was beginning to "harden" (Latour 1987), the monkey was increasingly able to support a conservation agenda.

Honor in the Eyes of the World

Science has always been articulated at a national as well as universal scale. This is true in a particular way for postcolonial nations where the hemispheric divide has produced an urgency in the pursuit of universal scientific knowledge. As a political prisoner on Buru Island, Pramoedya Anata Toer, Indonesia's most famous author, once wrote in a letter to his daughter Nen,

> Teilhard de Chardin is the greatest scientist of this century. . . . At a given moment in the seventeenth century, the light into the darkness of the preceding era was brought by Johann Kepler, namely in the field of astronomy; but, at present, into the darkness of the preceding era, the light is brought by Teilhard de Chardin, in the field of human evolution. . . . This is not a philosophy, this is almost a hundred percent science through which the truth is being proved. (Pramoedya Ananta Toer, cited in Mrazek 2002)

To understand how "the work of reason" is a "measure of things" in Indonesians' conservation biology, or how "a hundred percent science" is differentiated from a "philosophy," we must move beyond the articulation of interests (scientists, tourists, funders, and bureaucrats) brought together in the act of making the monkey.[3] How is conservation science productive of the Indonesian nation? And how does Indonesians' science constitute national subjectivities? In order to formulate the specificity of "Indonesian" as a modifier for conservation biology, or to think of the emergence of Togean nature as particular at the scale of the nation, we need to explore relations between science and nation in Indonesia in more general terms.

Recent anthropological work on national difference has proposed that something called "modernity" has alternative versions.[4] While maintaining a focus on difference within the modern world, I agree with those postcolonial and subaltern studies scholars who successfully avoid the sense that modernity's variations are a matter of citizens or nations choosing from among sets of "alternatives." For example, Partha Chatterjee (1993) has argued that the form of the postcolonial nation was never Benedict Anderson's Euro-"imagined community." Rather, as a national form, it was dependent on its status as successor to the nation imagined in Europe and the identitarian and conceptual struggles which that entailed. Similarly, Gyan Prakash (1999) describes India's "different" modernity as a "desire to institute a culturally specific community," yet he writes, "while successfully projecting the nation as an autonomous community, its imagination was overwhelmed by the history of the modern state." In other words, while the nation's affective attachments were formed in opposition to Europe, the state's own technologies of rule could not escape patterns established elsewhere.

Employing this postcolonial frame, both Gyan Prakash and Itty Abraham describe science in India as intimately articulated with the making of Indian modernities. Prakash (1999) describes how colonial science was "staged" as a sign of modernity, while Abraham (1998) reveals how the making of the Indian atomic bomb was part of India's struggle for national recognition. In each case, the work of these scholars emphasizes relations of science as forms of difference producing the postcolonial subject. It is impossible to distinguish Indian or EuroAmerican science from legacies of the nation-state, and this is no less true for Indonesian science.[5]

In Toer's novel, *This Earth of Mankind* (1996), we can see how science was present at the nation's birth. Pramoedya's story is structured around a narrator named Minke and his experiences in school. Minke's admiration for European science and technology leads him to make an unfavorable comparison between his tradition-bound father and his learned teachers. Yet Pramoedya's tale also questions the use and function of all this learnedness for colonizer and colonized alike. The Dutch in Indonesia are caught in a canonical colonial dilemma, namely what good is all that learning if in the end Europeans prove themselves uncivilized? Likewise, of what use is it for Indies natives to study the newfound science if, in the end, they are not allowed to apply their knowledge?[6] Only somewhere late in the story does Minke discover that the nickname he was given by his first Dutch primary school teacher had been his own mispronunciation of the word "monkey!"

While science and politics are inseparable, many scientists wish to see them divided on the basis of an idealized notion that good science is de-

Kalimantan, 1955. Classroom poster reads: "*Memperlindungi Alam Berarti Mendapat Penghormatan Dimata Dunia* (Protecting Nature Means to Receive Honor in the Eyes of the World)." J. Roberts/National Geographic Image Collection.

void of sentiment or opinion. Scientists in postcolonial Indonesia are forced to separate science from political sentiment for other reasons, as well. A biologist I knew in Indonesia once told me a story of how he learned to avoid politics. In 1965, when he was just a boy, his father was in the army. This was the year when the Indonesian army enflamed its citizens to brutally rid the countryside of suspected members of the Communist Party (*Partai Komunis Indonesia*, or PKI). His father was in the West Java battalion and he witnessed his father's compatriots round up suspected PKI members and gun them down in the open-air movie theater on the army base.[7] This history still informs the silent subjectivities of many Indonesians today and, as the mass killings progressed, my biologist friend witnessed the murders of his own school teachers. Between five hundred thousand and one million people died across Indonesia during this episode of recent Indonesian history (Cribb 1990).

With Suharto's rise to power in 1966 following on the heels of this violence, the nation was asked to reject the active political life that had existed under Sukarno; in place of "politics," the country would now pursue "development." Science, during the Suharto period, would come to carry the legitimacy of "progress" and "modernity" rather than the stigma of politics. Under these historical conditions, the seemingly apolitical nature of scientific inquiry was not an abstract philosophical position

on scientific objectivity. Educated Indonesian citizens were guided toward engineering, biology, and other fields in science and technology as ways to advance the nation through "apolitical" means. Although this does not indicate that Indonesian scientists might not also share a passion for the forms and mechanisms of biological nature, the appeal of biological science and species inventory must also be understood in the context of national trauma.

In the case of India, Prakash observes, science has been asked to "anchor the entire edifice of modern culture, identity, politics, and economy." The semiotic possibilities of science in Indonesia also include this "edifice" of modernity. While the Suharto government sought development as a panacea for politics, science was asked to complete the project of modernity by initiating the new Indonesian subjectivity governmental rationality proposed. We can see the way science and technology brought together the dream of technology and nation in mid-1990s Indonesia in the efforts of Dr. Bacharuddin Jusuf Habibie, Suharto's Minister of Technology. Habibie, who had been director of a German aerospace company in the 1970s, returned home at the invitation of Suharto to develop a "high-technology" economy in Indonesia, and Suharto granted Habibie unlimited resources to follow their mutual ambition of a scientifically modern nation. Out of this effort developed, most famously, the Airplane Industry of the Archipelago (*P.T. Industri Pesawat Terbang Nusantara*).

Habibie's idea of a high-tech economy was intended to create not only products for elite consumption but also new national subjects. Science and technology were imagined as central to an idealized Indonesian subjectivity that would transform the citizenry from top to bottom. Habibie once said: "The basis of any modern economy is in their capability of using their renewable human resources. The best renewable human resources are those human resources which are in a position to contribute to a product which uses a mixture of high-tech" (Head 1998). In the 1990s, the idea of "human resources" (*sumber daya manusia*) was a common, though of course indefinitely deferrable, way of describing the position of citizens within the nation.

While one might presume that this discussion of human resources would be limited to the elite center of Indonesia, marginalized peoples on Indonesia's periphery comprised the "other" in this national conversation. I encountered the term in the Togean Islands, for example, when a trader once explained to me, "the problem with these Togean people is their low human resource quality." While this rhetoric was surprising to me, a discourse of "human quality" was often used to distinguish Indonesia's cosmopolitan classes from its agrarian and fishing peoples. Togean Islanders were imagined, from an urban perspective, as backward, and

the "low-tech" technologies they used in their daily lives informed this representation.[8]

Biodiversity science in the Togean Islands shared with the industrial development of a "national airplane" and a "national car" (*mobil nasional*, or *mobnas*) the sense that its desired outcome was a technoscientific product belonging to the nation. Conservation biology in the Togean Islands was the science that would produce a "national park," although the sense of national patrimony always had to contend with the rhetoric of biodiversity's universal value. IFABS' conservation biology sutured international science literally onto the national landscape. It did so, however, within a set of transnational social relations still haunted by the "spectre of comparisons" (Anderson 1998) from Indonesia's colonial past. As a project of modernity-in-the-making, Indonesians' science entered upon territory the terms of which were known in advance and set somewhere else. As such, there was always the danger that Indonesians' science would be received as "repetition rather than re-presenting" (Bhabha 1994:88).

Beginning in the 1980s, new forms of activism in Indonesia mobilized the political neutrality of biological science and incorporated it into agendas of social and environmental transformation. For example, the Indonesian Forum for the Environment (Walhi), Indonesia's most prominent environmental organization, was able to advocate for land reform and worker's rights under the rubric of "environment" through the concept's affiliation with natural science. Walhi activists also occasionally employed the rubric of "biodiversity" during this period, although biodiversity was as much a means to legitimate their work with marginalized peoples as it was an expression of an abstract interest in extinction or systematics.[9] Both "environment" and "biodiversity" were deemed apolitical during the late Suharto period conveying a scientific valence both domestically and internationally.

While Walhi represents the environmental justice side of Suharto-era activism, IFABS, with its goal of advancing biological science, represented a somewhat more biocentric mode of social engagement. Togean primatology and conservation biology entangled Indonesian biologists in transnational scientific practices in which biology's distance from politics was explicit within scientific discourse itself. Rather than posit contentious political communities within the nation, or a variety of commitments to nature based on different and often conflicting sets of identities or rights (as Walhi sometimes did during this period), Togean conservation biology was conceived of—by Indonesian and EuroAmerican scientists alike—as producing politically uncontroversial outcomes.[10] While Togean nature emerged as locally specific through species inventory, Indonesian scientists gained generality as transnational and universal subjects who did not politically threaten the state. Moreover, IFABS scientists could be recognized for developing the nation's modern "human resources" at an unmarked cosmopolitan and universal scale, rather than for the localized "ethnoscience" through which foreign scientists and anthropologists had heretofore represented "Indonesian" ideas of nature.

The Social Life of Transnational Collaboration

That Indonesian conservation biology is now controlled by Indonesian scientists has everything to do with the existence of the Indonesian state. Unlike colonial-era natural historians, EuroAmerican scientists must receive research permits, work with scholarly counterparts, and demonstrate collegial respect in order to ply their trade in Indonesia. Many Indonesian scientists still choose to collaborate with foreign scientists in their research. The process of collaboration, though, is always haunted by the spectre of Indonesian scientific subalternity and the struggle for Indonesian scientists to "speak" (Spivak 1998) in transnational scholarly settings. In my work in Indonesia as a foreign scholar, I began to sense that tensions over EuroAmerican scientific hegemony often rested just beneath the surface.

My research began with my entrance into Indonesia through Jakarta and a visit to the offices of IFABS located near the campus of the University of Indonesia. Upon my arrival, Yakup startled me with the question: "Will you be working under your own name or under the name of Conservation International?" I answered that this was my own research project. "But why?" He told me that CI was claiming *they* had been doing biological research for several years in the Togeans, while the work was actually done by Yakup and his colleagues. The IFABS biologists were upset at their funder's failure to recognize them for their work. Because I had been initially introduced to the Togean project through a contact at CI, Yakup was concerned that I might share CI's outlook.

In addition to the sense that foreign scholars could easily overlook the work of their Indonesian colleagues, foreign scientists frequently caused difficulties for their Indonesian counterparts by not understanding local politics or cultural norms. Dr. Supriatna explained to me some of the problems he had experienced working with foreign experts. Once he was called from Jakarta by the local Sulawesi government to straighten out a disagreement among some of the foreign researchers he was sponsoring in a nature reserve. "This is not appropriate," he complained. And in a cautionary tale, he informed me that when some scholars did not turn in research reports to him, he had withheld their requests for visa extensions. He also warned me not to have any romantic affairs in the field, recounting the story of a woman anthropologist who had taken an Irianese lover, causing a scandal for her research sponsor.

Whereas in a U.S. academic setting difference among scholars is often understood as paradigm conflict, between Indonesian scientists and their foreign collaborators theoretical disagreements often produced tensions over the neocolonial relations of scientific collaboration itself. During one conversation at IFABS about my research, I described my project of comparing conservationist and Sama "ideas of nature." Yakup said what he wanted to know from me, however, was more about the "social structure" of the Togean people. He said: "In Java there is a social structure, which is strong and cannot be changed. Because Sama are used to being nomads, they are not used to being concentrated in such a tight group. This concentration is very dense. Therefore, there can be fighting and problems. These people will need to be dispersed."

Behind Yakup's invocation of "Javanese social structure" and the potential for "fighting and problems" among Togean people lay perspectives gained from his studies of biological anthropology and sociobiology. Behind my own interest in political rights and cultural representations of Togean people lay my training in political ecology and poststructuralist anthropology. As I was not willing to put my research to ends that this training had taught me to perceive as unjust or coercive, I explained to Yakup that I could only work to understand the perspectives of all sides. In this encounter, Yakup and I each were suspicious that our differences might lie at the level of nationality, not theory, however. In the context of transnational collaboration within the Togean project, North/South aspects of conflict were frequently emphasized over the abstraction of intellectual difference, and many tensions among foreign and Indonesian scientists were readily (and often mistakenly) viewed as "national" difference.[11]

Perhaps nothing caused more discomfort between Indonesian and foreign scholars than the question of research visas. Foreign scientists are required to have an official research permit before they do any studies in Indonesia. From the perspective of foreign scholars, the research visa can take months or years to get, sometimes requiring an expensive special trip to Indonesia to arrange. Since the visa can only be acquired in Jakarta, the process embeds the researcher in an endless network of bureaucracy from the capital to the village, and most importantly, once one is in Indonesia on a research visa it is impossible to leave without the written permission of the research sponsor. An American biologist I met once spent several days strapped to a board with a broken back waiting for permission from his sponsor before he could be airlifted to a hospital in Singapore.[12]

For their part, Indonesian scientists tend to believe that foreign scientists avoid collaborating with Indonesian scholars by entering their country on tourist visas. They are aware that EuroAmericans benefit professionally from their Indonesian research and believe foreign scholars will not share in these profits unless required to do so. By requiring foreign scientists to secure permissions through the Indonesian Institute of Sciences, the state attempts to ensure that foreign scientists share their knowledge and results with domestic scholars. In the 1990s the visa process was also a way for the national security bureaucracy to determine that foreign researchers were not engaging in politically sensitive research, or research understood as insulting to Indonesian sensibilities. Dr. Supriatna regularly confronted the issue of how to persuade foreign scientists to train students and field assistants, to share their research results, and to give due credit to Indonesian scientific contributions.[13]

In *Primate Visions*, Donna Haraway (1989) explores North-South scientific relations through the collaboration in 1975 of American primatologist Alison Jolly and Etienne Rakotomaria, Director of Scientific and Technical Research in the Malagasy Republic, as evidence of the decolonizing process occurring in transnational science. This process is especially salient for conservation biology, which requires the literal space of postcolonial nations to carry out its work.[14] Haraway cites Rakotomaria, "Scientists will only be allowed to work here if they arrange reciprocal benefits for Malagasy colleagues. The people in this room know that Malagasy nature is a world heritage. We are not sure that others realize it is our heritage." Haraway identifies this as a beginning of "Malagasy-controlled reconstruction of what nature and conservation must mean for national, ecological, and cultural survival," and she asks a question equally salient for Indonesia: "What might a postcolonial reinvention of nature be like?" (Haraway 1989:207).

Camp Uemata

At Camp Uemata, I experienced biological field science with the IFABS scientists firsthand. In 1996 and 1997, Uemata was a busy "center of calculation" where data on Togean nature was not only compiled but also made mobile, stable, and combinable (Latour 1987). During the course of the year I met many Indonesian scientists, including several primatologists, a marine biologist, a forest ecologist, a geographer, an economist, and an ornithologist. I also met foreign researchers, including an expert on the coconut palm, a graduate student studying public health policy, a political ecologist, a biological anthropologist, and the members of two Earthwatch expeditions. Another cadre of Indonesian experts on tourism

and development worked to translate biological mandates into social programs. And finally, Togean Island people assisted scientists with conservation and research activities. Both experts and expertise were outcomes of Togean field science.

One night I went into the forest with an American primatologist who was conducting a study of the Togean tarsier. The tarsier is one of the world's smallest primates and the species status of the Togean Island tarsier, like that of the Togean macaque, also hung in the balance. Participating in capturing the tarsiers that day were Dr. Supriatna, Pak Arif (a field assistant to the American scientist), and several American volunteers from an organization named Earthwatch. We climbed up to a dead ficus, which the scientists suspected was the sleeping tree of a family of tarsiers. Arif had previously set up mist nets surrounding the tree to catch the nocturnal tarsiers when they came out at dusk. I sat near a volunteer who took my photograph saying: "Anthropologist at work in the jungle!"

Suddenly, we heard a chirp. Then, immediately, a whistle. This was the duet call of a male-female pair that the American primatologist was recording to use in a behavioral analysis of tarsier speciation. It didn't take long before, zip!, I saw the dim outline of a tarsier leap from one leaf clump to another and it was caught in our mist net. Looking into the net with our flashlights, Dr. Supriatna asked the volunteers if they had ever seen a tarsier before. He believed this was a new species and the largest kind of tarsier, weighing 150 grams. He explained, "There are four species of tarsiers in Kalimantan and others in Sulawesi each having color variations. The tarsier won't live in captivity. They don't eat because of stress," he added. "Are they endangered?" the volunteer with the camera asked him. "No, not endangered, but protected," Supriatna replied.

Dr. Supriatna had recommended Arif to the American scientist as an expert on Sulawesi primates and on the techniques of primatological field work. Pak Arif lived near Tangkoko National Park in North Sulawesi, which has populations of both macaques and tarsiers. He had made his career working with both Indonesian and foreign scientists, and he attributed his entry into the scientific and conservation worlds to the wide variety of primate researchers he had met there. Arif assisted them with building traps, setting nets, taking blood samples, and taking care of the physical aspects of primate field study that scientists generally find difficult. Arif also had a detailed understanding of primate behavior. Although he had only an elementary school education, this was not an obstacle to demonstrating his knowledge, and he was well regarded in the world of Indonesian primate research.

Dr. Supriatna promoted all his Indonesian staff, whether scientist or assistant, into positions of authority at Camp Uemata. He explained to me, "Arif is the best tarsier man in the country." Another time he invited

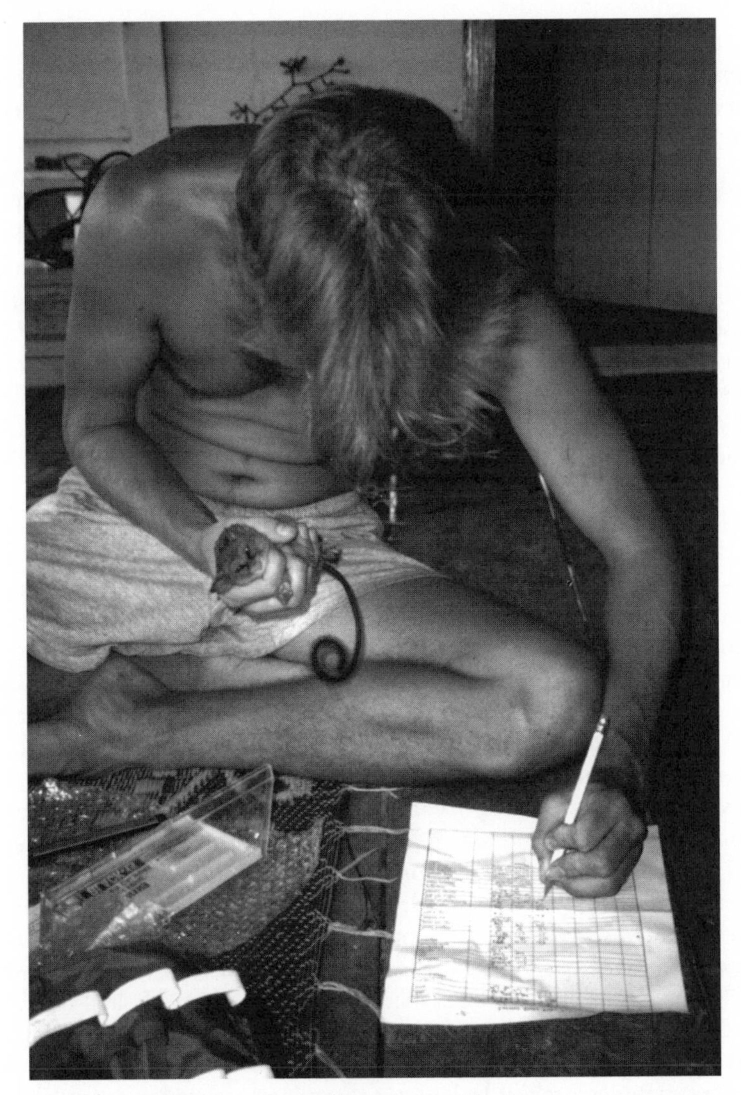

Measuring a Tarsier at Camp Uemata, by Celia Lowe.

Pak Ahmad, the Ranger, to explain to some camp visitors his experience working with the biologists. Ahmad described how he had assisted Budi with his study of Malenge's trees: "In 1992, Budi first came to build a camp here. We built a trail through the forest and we made a 'bell' and then a 'plot.' A plot is an area 50 meters square where we counted the number and size of trees. We also counted the six types of fruits and the seven types of leaves. This is what is called 'analysis.' "

While Dr. Supriatna attempted to situate his Indonesian colleagues as experts, sometimes his efforts would fail dramatically. On one occasion, a EuroAmerican scientist complained loudly about his food saying, "It seems whenever I'm eating with a group of Europeans and a bunch of foreigners, I can't make them understand not to use so much chili pepper." Budi got up and left the table abruptly: who, after all, was foreign here? Or once, when an Indonesian biologist showed up late for a field excursion, I heard a white scientist proclaim, "that's why they're still beating on drums." Most frustrating for this group of elite Indonesians was that foreign visitors to the camp were sometimes unable to make a distinction between Indonesian scientists and Togean villagers. The distinction between cosmopolitan scientists and rural Togean people, however, was one of the important social distinctions at stake for Indonesian experts in their scientific production of nature, and the field station was a place where "modern" and "traditional" identities were continually reinscribed.

At Uemata, Indonesian scientists were also, at times, susceptible to representing Togean people as "unknowing," and Togean knowledges could be overlooked in the scientific effort to reveal biological nature. My Sama friend Isra once asked me if I believed we are descended from monkeys. I was a bit noncommittal on the subject. The 35 million years that separates me from the Togean macaque *does* seem rather a long time to recon kinship and, as Franklin and McKinnon (2001:5) write, "Kinship provides a useful example of naturalization as knowledge because of the way in which kin ties are seen to be constituted out of primordial natural facts."[15] Isra asked me about primate kinship because he himself had been brought to the camp by IFABS in the hopes that he would internalize their conservation values. He learned many things about the conservation project while he was there, including the urgency of preserving the Togean monkey. One explanation he was given of its value was that it is our "ancestor." But, he told me, he did not believe humans are related to animals at all. "God looks at humans differently than animals," Isra said.

Biologists' desires to revise Togean peoples' views of nature were reflected in the way they had reworked physical space at Camp Uemata. Uemata was built on an uninhabited beach, but the reef in front of it had previously been a sea cucumber collecting and fishing ground for Sama people from nearby Pulo Papan. Indeed, the hill behind the camp was somebody's garden plot. Uemata itself was named after the tiny spring—not much bigger than tears—that had flowed onto the sand at the end of the beach before it was turned into the camp's shower facility. A garbage pit at one end of the beach now gathered trash and flies. The biologists' idealist supplanting of Togean peoples' natures was physically replicated in their appropriation of the land- and marine-scapes of the campsite.

Pak Ahmad, who was positioned as a culture broker between the scientists and his Malenge Island neighbors, knew that the Togean macaque competed with Malenge people who needed the space to grow coconuts on the island. One day, as I was speaking to some women who were coring coconuts, Ahmad teasingly called them "monkeys." I joined in the play by asking if they were a "protected" species. Ahmad responded, "These are more protected than the ones in the forest." From his perspective, if you own coconuts, even if you are sick or old, you have something of value. If you want to go to the store and ask for credit, the owner will give it to you. He will pay your taxes for you, he will do almost anything to help. For Ahmad, the scientists and their idea of biodiverse nature threatened this livelihood and promised to make Togean monkeys more significant than Togean people.

Scientists and Togean people disputed the identity and meaning of Togean nature and, even as the scientists hoped that Togean people would come to appreciate their biodiversity perspective, biological and Togean cultures of nature could not always be reconciled. Isra claimed the monkeys did not even *come* from Malenge Island. He told me how they had arrived on a Portuguese ship that had anchored near Malenge a long time ago. The ship's crew put down a board to the island and the monkeys escaped. So while the scientists were searching for the origins of the monkey in evolutionary time, Malenge people described equally exotic origins of the monkey from historical memory.

M. Togeanus Becomes a Dubious Name

At the end of Supriatna and Froehlich's 1996 article suggesting the likelihood of the Togean macaque's species status, the authors mentioned a story similar to the one I had heard from Isra: "Recent fieldwork supports local traditions that the monkeys of Malenge Island were artificially transported there about 1920 from near Tanjung Api . . . they occur nowhere else in the Togian archipelago." This story of the historical origins of the Togean macaque sits uneasily in the same article alongside the authors' invocation of island biogeography, which was intended to indicate the possibility of the monkey's arrival by a prehistoric land bridge.

Nonetheless, Isra's "local knowledge" could not overwhelm the momentum of the primate research or the conservation effort. Dr. Supriatna and his students sent blood samples of the Togean macaque to Columbia University in New York for analysis, and he and Froehlich conducted field research on the mainland to look for related monkeys. Over nineteen days in June 1995, and twelve days in January 1996, they sampled ninety wild and pet "Togean-like" monkeys from around the Balantak

peninsula. They sedated the monkeys by injecting them with Ketamine HCl using hand-held syringes or blow-guns and injection darts, and they gathered fingerprints on clear tape covered with graphite powder. The macaques were examined for blood, dermatoglyphic volar prints from both hands and feet, and morphometric and pelage data. In order to assess the health of the Togean macaque population, they took bilateral cheek tooth measurements from the 1939 Menden collection and assessed asymmetry scores to look for signs of stress on the Togean monkeys (Froehlich et al. 1998).

In August of 1997 Supriatna, Froehlich, and three colleagues submitted their reanalysis for publication in the journal *Tropical Biodiversity.* This time there was no mention of prehistoric Togean biogeography, and the story of the Togean macaque's historical arrival on Malenge Island was covered in specific detail:

> Oral traditions on Malenge and the mainland tell of a few animals being transported to this island in the 1920s, supposedly by a Swiss man named Sibley in charge of German copra plantations between Tanjung Api and Bunta starting in 1918 (Pak A.Ex: Palali [Luwuk Historian], pers. comm.). As an amateur naturalist, Sibley sailed often around Teluk Tomini in his orange boat. Residents of Malenge also tell of more recent, but failed attempts to establish their monkeys on neighboring islands of the Togeans. (1998:168)

In their new article, the scientists conclude that the Malenge population from Menden's collection includes genes from female *M. tonkeana* and male Balantak monkeys. "This hypothesis is supported by the emphatic testimony of Pak Amir, a very knowledgeable forest guide on Malenge Island," they wrote (177), since Amir had never seen traits indicative of the Balantak female. At the same time, the allure of taxonomic discovery remained alive, and they determined that the Balantak peninsula was home to a different, previously undescribed, species of macaque that they nicknamed "the Balan of Balantak" in the title of their article.

At this point, the scientific research findings failed to sustain the independent species status of *M. togeanus*, and the macaque was concluded to be a "bottle-necked hybrid swarm" or, as another primatologist described it to me, "feral." Their new work reversed the suppositions of the 1996 article with the following description, "The small Togean population represents an unnatural hybrid swarm, bottlenecked and inbred after its founding mostly by female Tonkean and male Balantak genes." Furthermore, "The Togean sample is . . . not representative metrically, genetically, or historically of the Balantak monkey on the Sulawesi peninsula." The new field data do support the idea that the Balantak primate is a valid

new species, while "*M. togeanus* is a *nomen dubium* for the population on the mainland peninsula" (180).

So what would become of this monkey as a raison d'etre for Togean biodiversity conservation?

From Dubious Name to Endemic Species

The status "dubious name" might have been the end of things for *M. togeanus*, but it was not. In the Balantak article, the authors argued again for the conservation value of the Togean primates: "It would seem advisable to declare the name *Macaca togeanus* a *nomen dubium*, or perhaps to reserve the name for a unique hybrid subspecies of *M. tonkeana* in order to facilitate its protection for the great potential it offers in the study of inbreeding and of hybridization between primate species" (Froehlich et al. 1998:180).

By 1997, although the Togean macaque had come full circle from its recognition as a "new form" to a "dubious name," the existence of an "endemic" primate species still had utility for Togean Island conservation. As Supriatna had explained, "charismatic animals allow people to want to save the environment." At the International Tropical Marine Ecosystems Management Symposium (ITMEMS) held in Port Townsend, Australia in 1998, well after *M. togeanus* had become a "bottlenecked hybrid swarm," Dr. Supriatna and an Indonesian colleague presented a paper in which they argued: "Almost sixty percent of the land area of the Togeans is covered in tropical forest that supports an impressive array of local and Sulawesi endemic species including: the Togean macaque (*Macaca togeanus*)—a primate only recently described in 1996 (Froehlich and Supriatna 1996); the Togean lizard (*Varanus salvator togeanus*); the babirusa or 'pig deer' (*Babyrousa babyrussa togeanus*); and the Togean Tarsier (*Tarsius togeanus*) (Surjadi and Supriatna 1998:281)."

In referring only to his co-authored 1996 article where Dr. Supriatna had argued for the macaque's endemic status, it is clear that the outcomes of conservation biology and primatology alone were not a sufficient basis upon which the project might reverse its course. "Science in action" (Latour 1987), in this case, contained the momentum of the scientists' investments and interests in the Togean Islands as a protected area and a biodiverse locale. Already they had committed too much to reverse the project's course based on scientific findings alone.

The status of the Togean macaque as "endemic" was a strategic essentialism constituting a means to several ends. First, the science of "making the monkey" established a way to legitimate the Togean conservation project within the larger sphere of Indonesian state control. It also had

secured a place for Indonesian science within transnational field biology and primatology in which the *presence* of scientific practice was as least as important as its outcomes.[16] Third, it committed CI to funding Togean Island conservation. And finally it had succeeded in transforming IFABS into a new institution able to promote biological science in Indonesia.

These social facts were sufficient to secure the status of *M. togeanus* as unique and worth conserving more solidly than dermatoglyphics or pelage data could, and it would not prove possible to return Malenge Island over to Togean peoples' coconut farming or to local meanings of nature on the basis of any scientific results. Once a "species" has entered into the realm of biodiversity calculation, it hardly ever emerges again as unworthy of protection; transnational biodiversity ideologies inhibit such reversals. This, of course, has implications for the people who live in proximity to biodiverse nature without sharing its logics. As Ahmad had said to me: "We can't run to the land; we can't run to the sea; where can we run?"

But this is a world only partially constituted by Indonesian scientists and their own social and scientific initiatives. In her keynote address at ITMEMS, Nancy Foster (1998) of the United States National Oceanic and Atmospheric Association, argued for moving away from species-driven biodiversity conservation. Instead she proposed "an ecosystems approach to management" as the appropriate scale for both scientific analysis and political advocacy. This was a moment when conservation was beginning to shift its rationale from "species" to "ecoregions." Dr. Supriatna's doubly anachronistic promotion of the endemic macaque at the same conference reflects postcolonial relations of knowledge and knowledge-making in which those at the margins of scientific production are allowed to contribute data, while those in centers provide the theories through which scientific reason will be known. We can see this in Supriatna's own primatology where his advisor, Jeffrey Froehlich, is recognized for his invention of the techniques for dermatoglyphic analysis, while Supriatna's research into speciation and hybridity constitutes a case analysis.

In this sense, the creations of Indonesian scientists, like the Indian scientists described by subaltern studies scholars, are recognized only through struggle and insistence. To imagine "nonderivative" thought within Supriatna's scientific project, however, we need only look at his scholarship. Specifically, his insistence on recognition of Indonesians' contributions to the scientific project encompasses its own theory of standpoint. Supriatna's scholarly writings contain continual reference to the knowledge of Sulawesi and Togean peoples, and he makes their perspectives significant for understanding the facticity of the Togean macaque itself. From the first he includes the story of the macaque's arrival

on Malenge by ship. In his later article he cites Pak A., "a Luwuk historian," and Pak Amir, "a very knowledgeable forest guide," for their knowledge of the macaque. His citational practices consistently allow for Indonesians to speak as experts—be they undergraduate students or local oral historians—in situations where EuroAmericans might otherwise overlook these contributions.

M. Togeanus as Keystone Species

The Togean macaque, "*M. togeanus*," was a metaphorical "keystone species" for Togean conservation biology. In its biological sense, a keystone species holds together an ecosystem and, in the Togean Islands, the macaque held together the social life of biodiverse nature. The monkey drew together threads connecting Indonesian biologists to the ideas of biodiversity extending across nations in the 1990s, while the scientists were also positioned within domestic ideas of nature and nation. In this way, Indonesian biologists became subjects of a universalizing project of scientific inquiry while simultaneously participating in the making of an Indonesian modernity that was historically and geographically distinct. In this double move, Indonesian modernity is not an "alternative" to something more central or originary. Indonesian conservation biology produces scientific reason in a manner that should be familiar from the many projects of knowledge production described by science studies scholars, while it is no less entangled with those issues of origin and identity outlined by theorists of postcoloniality.

Indonesians' conservation biology gave the IFABS biologists an elite location within the nation, and this social position, in turn, gave them a strong voice in relation to Togean Island peoples who could assist scientists in the production of biodiverse nature but could not overturn the biodiversity paradigm. By imagining Budi, Yakup, Dr. Supriatna, and the other Indonesian biologists as occupying that space of the middle—between, on the one hand, international conservation and science and, on the other, Togean Island people—the postcolonial status of Indonesians' science becomes apparent. The story of making the monkey indicates the complexities of living in a postcolonial world where Indonesian scientists are both elite (within the nation) and subaltern (within transnational science) at precisely the same moments.

Chapter Two

THE SOCIAL TURN

> A growing number of international organizations have chosen to focus on Indonesia including ODA, USAID, JICA, WWF, Birdlife International, Wetlands International, WCS, FFI, WPTI, CI. In their hearts glow two effective combinations of approaches: site based (e.g., adopt a park), and empowerment of local communities and development of small enterprises. . . . IFABS endeavors in Togean were started in 1992 by Dr. Jatna Supriatna who focused on ecology and conservation of the Togean Macaque. In 1993 a greater effort was spent to make inventories of Togean's marine and terrestrial biodiversity. Needless to say, maybe, inventory is the base foundation on which biological management is built. IFABS, under the leadership of all IFABS Directors, realized that encouraging local community participation is the next prerequisite for successful conservation and has accordingly to strive to turn the local communities to be the main beneficiary of conservation. The wheel has turned and begun to run now.
>
> —Didi M. Indrawan, *Notes on Togean Conservation*

AT CAMP UEMATA I met a scientist named Laksmi, an IFABS biologist who maintained a strong interest in Togean people and who was concerned with how regimes of conservation and development could become coercive. Similar to the way many anthropologists have been attracted to their field, Laksmi's interest in Indonesia's fishers and farmers developed through an early encounter with representations of Native Americans.[1] When she was young, Laksmi had read the works of German novelist Karl May translated into Indonesian. Unlike the depictions of the valiant cowboy and treacherous Indian of American film and literature, May painted a picture of a "noble" Indian under siege by the savage cowboy.

Laksmi sympathized with Winnetou, the native protagonist in May's books, and she began to compare him in her mind with Indonesia's many marginalized ethnic groups. Her desire to equate a fictional representation of Native Americans with an equally imagined understanding of Indonesia's marginal peoples was facilitated by the Indonesian state's own framing of permissible forms of ethnic difference of the "song and dance variety" (Li 1999). Many scholars of Indonesia have written about the state's

instrumental use of "culture," and of its efforts to aestheticize and depoliticize ethnicity during the New Order period (Errington 1998; Volkman 1990; Pemberton 1994; Hefner 1990). But unlike these official stagings of culture, Laksmi's imagination was also informed by a deep dislike for inequality.

As Laksmi entered the world of biological science, she continued to imagine the plight of many Indonesian peoples through her sympathetic reading of Winnetou. Her imagination took on concrete form during her first trip to the United States when, identified as one of Indonesia's future leaders, Laksmi was invited by the United States Information Agency to tour the country. Able to travel anywhere, she requested to see a place where Native Americans were managing their own natural resources. She was interested to learn if there were resource conflicts between local people and national parks in the United States, or if only Indonesia had this experience. Laksmi was taken to Menomonee, Wisconsin, and to the Adirondack mountains of New York where she learned that, indeed, such conflicts exist.

I discovered, myself, the ability of the Native American (and the Australian Aboriginal) to stand in for the marginalized ethnicized Indonesian (especially outside of Java, Bali, and Sumatra) when I was in Palu, the capital of central Sulawesi. When I presented my research proposal at the Department of Social and Political Affairs I was told, "you have your Indians, and we have our Bajau." Yet, among some Indonesian biologists, the Native American experience was also a guide to what should not happen in Indonesia. They were aware of how Native Americans had been expelled from park lands by the United States Army to create spaces of Arcadian wilderness, and they used this as evidence Indonesia needed to follow a different path. Scientists like Laksmi wanted to work to incorporate people into a national park concept that was more appropriate for their country.[2]

Conservation biology brings biologists into intimate contact with the question of what to make of the peoples inhabiting the "natural" places they study. My friend Hari developed a new understanding of the problem when, as a student at the University of Indonesia, he was invited by Dr. Supriatna to join the Sulawesi Primate Project. In studying the macaques of South Sulawesi for his undergraduate thesis, Hari began to appreciate for the first time the lives of the people of rural Sulawesi and the everyday problems they faced. In the case of the Sulawesi macaque, the issue that brought together Sulawesi people, monkeys, and conservation was known as the "crop raiding problem" (Supriadi and Akbar 1997). Sulawesi farmers thought of the macaque as a pest, since the monkeys took coconuts from trees and crops from fields. Farmers were trapping and

poisoning the monkeys to protect their crops and, as farmlands expanded, conflicts between farmers and monkeys had become more intense.

While some biologists viewed Sulawesi farmers primarily as a threat to the monkey, the crop-raiding problem radicalized Hari's thinking. Through his scientific interest in protecting monkeys, Hari came into contact with Sulawesi people as objects of scientific study and political advocacy. On the one hand, he wanted to know how humans were affecting the survival of macaque populations. On the other, he wondered if the crop-raiding problem might be resolved in a way that was beneficial to both people and monkeys. For this, biology alone could not help him. Conflicts between people and primates raise larger issues of access to resources and of how Sulawesi peoples understand the landscapes around them. Hari's questions also interfaced with another issue important from the perspective of Indonesian developmentalist discourse, namely, could Sulawesi farmers' lives be improved?

Laksmi and Hari brought these experiences to their understanding of the relationship between Togean nature and Togean people and, thus, the people of the Togean Islands became objects of study. In the early part of the 1990s, Laksmi and Hari explored Togean peoples' knowledges through the lens of "ethnobiology," combining the biological study of "ecology, vegetation, flora, [and] land and sea fauna," with the anthropological study of "knowledge systems concerning the technological uses of flora and marine fauna" (Yuliati, et. al. 1994). Then, in the mid-1990s, the form of knowledge the scientists sought changed to adapt more closely to the needs of political advocacy, and IFABS began to develop the idea of "spatial planning" (*penataan ruang*). In this, the biologists broadened their definition of knowledge to examine Togean peoples' ideas of space and place.

Indigenous knowledge came to be seen by IFABS scientists as an important supplement to the biological knowledges produced in species inventory. Framing Togean people as "indigenous," and therefore possessing special knowledges and abilities, was characteristic of a transnational social turn in biodiversity conservation. If "biodiverse" was the form scientifically produced nature acquired through Togean species inventory, the "indigenous" was the form that culturally produced society simultaneously assumed. Through ethnobiology and participatory spatial planning, the scientists began to collaborate with Togean people—understanding that conservation could not move forward without their participation. In multiple and sometimes contradictory ways, indigenaeity was the analytic through which Togean people became manageable and knowable subjects of the Indonesian state, and through which the cosmopolitanism of Togean peoples could be denied. Indigenaeity was

also, however, key to Togean peoples' discursive emergence as rights-bearing citizens of the Indonesian nation.

How did such a seemingly solid object as "indigenous knowledge" emerge as a solution to the controversies of Indonesians' biodiversity conservation, and why was "indigenaeity" the form assumed by the human in this work? Bruno Latour suggests we track how observers of nature (and culture) move in space and time, how their inscriptions are enhanced, and how their networks are extended. Another method for inquiring into the specificity of knowledge-objects in the present is to look at how relationships between nature and the human have been problematized in the past. Colonial era nature-making also produced observations of how Indies and Indonesian peoples related to the world of plants and animals, and produced inscriptions bearing representations of Indies peoples.

The Togean Islands seem far away from cosmopolitan interests, yet the islands have been a colonial and postcolonial entrepôt for discovering, enumerating, describing, and cataloguing nature's contents since the mid-nineteenth century. The archipelago first became a site for nature-making beginning with von Rosenburg's travels in the Gulf of Tomini in the 1860s, Adolf Bernard Meyer's survey of birds in the 1870s, J.H.F. Umbgrove's survey of corals in the 1920s, and J. J. Menden's collection of macaque skins in 1939. If we slightly expand our geographical purview to encompass the eastern half of the Indonesian archipelago, we can push our comparison back even further in time. In the seventeenth century, Georgius Everhardus Rumphius devoted his life to documenting nature's plant, animal, and mineral contents in the nearby Maluku Islands. None of these early projects framed Indies peoples as "indigenous" or accorded their knowledges particularity or peculiarity on that basis. The work of these early natural historians reveals the arbitrariness of indigenaeity as both problem and solution to the relationship between nature and the human.

The Nautilus and the *Mestika*

In the history of systematic inventory of life forms in the Malay world, Rumphius's work is the earliest, and certainly the most monumental. Arriving in the Indies from Germany in 1653 at the age of twenty-six, he spent the next fifty years until his death in the Moluccan capital Amboina developing an account of Indies "curiosities." Although he was officially an engineer, ensign, and junior merchant for the Dutch East Indies Company (VOC), his calling was economic botany and he spent his time collecting, describing, and illustrating the natural wonders of what he called

the "Water Indies." His famous compendium, *The Ambonese Curiosity Cabinet* (1999[1705]), is an enormous volume divided into three books: *Dealing With the Soft Shellfish*, *Dealing With the Hard Shellfish*, and *Dealing With the Minerals, Stones, and Other Rare Things*. His other great work, *The Ambonese Herbal* (1981 [1741]), comprised of 1,662 folio pages divided into 876 chapters and illustrated with 695 plates, describes 1,200 species of plants (Beekman 1999:lxxx). Rather than the now familiar "flora and fauna," Rumphius is known for his record of Moluccan plants, sea creatures, and stones.

In the seventeenth century, the Molucca Islands (now Maluku) were the center of natural endemism for two botanical wonders: the clove and the nutmeg. The VOC concentrated its entire military and administrative strength on maintaining its monopoly over the spice trade and wanted to find other similarly valuable commodities. To justify his work to his employers, Rumphius's studies had to be "of use and service to those who live in the East Indies" (lxxxi).[3] But Rumphius also intended his discoveries to be "diverting to Lovers of Nature," and this is clearly where his passion lay. At the age of forty-two, Rumphius went completely blind from "black cataracts" (glaucoma) yet, despite this impairment, continued documenting and describing nature's objects with the help of assistants until his death in 1702.

The problem Rumphius addressed was similar to that facing conservation biologists today, namely, What forms does the natural world contain? His method for determining nature's contents, however, was an entirely different one. For Rumphius the answer came not only through Malay peoples' knowledge, but through engaging Malay people's *thought*, and Rumphius often aligned himself with that thought.[4] Not only did the people of Ambon help Rumphius collect and catalogue, they also transformed the structure of his classifications and helped to shape his interpretations. In *The Ambonese Herbal*, Rumphius followed the pattern laid down by Aristotle, that is, dividing botanical descriptions into three main groups: trees, herbs, and shrubs. Yet in outlining the Indies flora, Rumphius equally follows Malay categories, dividing plants into "edible, fruit-bearing, aromatic, and wild trees"; "all sorts of shrubs which stand upright"; and "such Shrubs which cannot stand upright on their own, but which creep along with a long and slender trunk, or wind around other trees, the like we call in the Indies Forest-ropes, or Taly-Outang in Malay" (lxxx).

Rumphius related what Malay people told him about a thing, how it was used, and how the wonder came to arrive in his possession. The Nautilus, for instance, came to him through the agency of a fisherman and a sea eagle:

> I should mention a rare event here. A Sea Eagle (Haliaetos), being a bird that constantly hunts at sea, took such a Nautilus, while it was floating in the Sea, and bore it aloft, but while his business was with the fish, and since he did not care about it as a curiosity, he struck his claws mostly into the fish, wherefore the shell came to fall out of his claws, and by rare fortune, it fell on a small spot of sand between rocks in such a way, that nothing was broken off, except for a small corner of the foremost edge; and a fisherman who was wandering thereabouts, quickly picked it up and brought it to me. (Rumphius 1999 [1705]:94)

Rumphius's pattern was to describe the plant, creature, tree, or stone in detail and then follow with its name in various languages. Although he wrote in Dutch, European and antique names were not yet transcendent in Rumphius's taxonomy. For example, he described the Nautilus as a biologically hybrid creature having the mixed nature of fish and shellfish and wrote, "their name in Latin is *Nautilus tenuis* & *legitimus* and its little Boat, *Carina Nautili*. In Dutch *Doekhuyven*, because of the many folds, In Malay, *Ruma Gorita*, that is *Domuncula Polypi*, just as the Greeks called it *Ovum Polypi*. In Ambonese *Kika Wawutia*" (Ibid). Rumphius could read and write Arabic script; read, write, and speak German, Dutch, Malay, and Latin; speak Portuguese and Ambonese; and may have known some Chinese and Macassarese. He also had a limited knowledge of Hebrew and Greek (Beekman 1999:lvx).

While Rumphius occasionally took European categories and searched out their Malay equivalents (as in his title, "The Malay Names For Some Precious Stones"), he was just as likely to allow Malay categories to illuminate nature's contents. One Malay taxa that worked its way into Rumphius's compendium was the *mestica*. The mestica was:

> any little stone that one finds in a plant, in wood or in other stones, in a way that is not nature's wont, and that was produced by the same, not from a corruption that would afflict an animal, but purely from an excess of good food, which has attracted a stone like sap from the external food. Hence we do not include such stones which happen to get inside animals or plants or which produce any disease in the same, such as those that are found in the stomach, Urine, gall bladder. . . . But a Mestica must be a hard little stone, smooth, seldom or never transparent, produced in places where nature would not have engendered it herself, such as in flesh, brains, fat, and on the bones of animals, in pure wood and in the fruits of certain plants. (Rumphius 1999 [1705]:327)

One type of *mestica* was the "tiger stone," found in the head of the Sumatran tiger. Rumphius wrote, "one could consider them [tiger stones] the heads of people, horses, and other animals, which this Murderer might

well have devoured when he was alive, and these images had been impressed on his brains and that stone by his imagination" (Ibid., 329). Rumphius's nature was animated through the agency of people, plants, and animals. It was a nature where eagles delivered curious objects into fishers' hands, and where murderous tigers carried images of their victims literally in their heads. The question for Rumphius was where those powers lay, not whether they existed: "It is true that the common people tell of many strange events, that they have seen people, who could not be killed with any kind of weapon, until one or more of those little stones had been cut out of their bodies where the same had been pushed in; our people have confronted such cabalized people in war, and these things have been said by our own officers, so that I do not care to contradict the same."

Rumphius was not a Romantic (a movement that did not yet exist) about Malay peoples or the natures they introduced him to; many hidden powers he considered "fancy and superstition" (327). Still, Rumphius's nature-making did not open up a civilizational divide between himself and those Malay people whose company he kept. The Ambonese people did not emerge as a special case of humanity possessing a particular kind of knowledge. They were, instead, people who assisted Rumphius with thought. Rumphius asked of the people around him what objects were found there and how they operated, not what did Ambonese people "believe" to exist. As such, the "indigenous" could form neither problem nor solution in his work.

The *Madreporaria* and the Lorikeets

The most important precursor in the direct lineage of Togean biodiversity conservation was J.H.F. Umbgrove's study of coral reef morphology and speciation in 1928. As a scientist aboard the *Eridanus*, a hydrographic vessel of the Netherlands East Indies Navy, Umbgrove inventoried the *madreporaria* (a genus of hard coral) of the Togean Islands (Umbgrove 1930, 1939). He identified six new species and one new variety, collecting seventy-six species in all. In his analysis, he compared the Togean corals with those from Cocos-Keeling, Batavia (Jakarta), Amboina (Ambon), Murray Island, and Samoa, thereby elevating the Togean Islands to prominence in the history of natural history. His collection of Togean corals was packed by his Javanese aide, Jusuf, and shipped to the *Rijksmuseum van Natuurlijke Historie* at Leiden and the Museum of the Geological Survey at Bandung, Java.

"Where, one asks, is everybody?," muses Mary-Louise Pratt (1992:51) in her discussion of colonial natural history. One may ask the same question of Umbgrove's Togean Island collecting adventures. All we learn

from him is the name of his Javanese assistant, Jusuf, and of Jusuf's menial task. We know of no knowledge provided by this assistant and, moreover, Umbgrove never mentions Togean people. In Umbgrove's form of natural science, Malay peoples could not be experts on Indies nature. Their ways of knowing were neither necessary to his work, nor were they worth remarking upon. This was not precisely the case in the earliest recorded study of Togean life forms, that of A. B. Meyer's collection of birds in the 1870s, where both birds and Malay people took on a form more substantial than mere absence.

On the one hand were the Celebes birds. Meyer describes his wanderings over the countryside shooting nearly every bird in sight or, if birds flocked in great numbers, his desire to bag at least several of every kind, age, and sex. The dead birds were then delicately dried or preserved in alcohol or sprits. "If the dead bird is pressed on the belly," Meyer wrote of the lorikeet, "the same cry, *krok krok*, can be made, as from the living. I found this the case even in specimens preserved in spirits of wine" (Meyer 1879: 67). Bundles and bottles of tiny corpses were then shipped off to England to form part of the accumulating collections that would define an ever more universal and abstracted nature.[5]

On the other hand were Celebes peoples. In Meyer's depictions of native scientific incompetence, they acted as foil for Meyer's scientific appreciation of nature. "The natives do not distinguish this species from *P. platurus* by a special name; generally they are not strict observers of nature, at least not strict in our scientific sense," Meyer wrote (51). Meyer referred to all Celebes peoples as "Alfurese," the colonial name for the many peoples of the Celebes interior, and he found cultural and linguistic diversity messy and untrustworthy. "Native names often change from village to village, as the native language generally does in these parts of the East." Likewise he viewed their classifications as unreliable:

> Neither can I agree that the natives recognize them [*T. muelleri* and *T. ornatus*] as two birds. Even if they did, I should not attribute much value to such a statement, as generally the natives of Celebes know but little of their fauna, and answer a question as they think will most please the questioner. Nevertheless, were *T. muelleri* a species often kept in captivity by the natives, as is *Trichoglossus ornatus*, I could perhaps trust them; but this is not the case, *T. muelleri* rarely being seen in captivity, on account of its unamiable character, at least in this state. (47)

People were warlike while nature was peaceful in Meyer's accounts. "The village of Posso [to the South of the Togean Islands] is a fort," he writes, "the natives being almost constantly at war with their neighbors, and even when I was among them; they keep the skulls of their enemies in a hut in the middle of the village. There is a group of large trees between

the houses; and at noon I saw some white *Cacatuae* sleeping in the foliage, a striking contrast of peace in nature to war among mankind. I left those pretty birds undisturbed in their high resting-places" (46).

Patricia Spyer's account of "paradise's birds" on the eastern Indonesian island of Aru describes the cultural economy of birds in the nineteenth-century Indies (Spyer 2000:41–65). Bird feathers were important for hat making in Europe, and Spyer writes of a "naturalist's wife" who observed the export of 2,000 orange-feathered birds of paradise, 800 kingbirds, and various other birds from the port of Makassar in South Celebes in the 1880s. Between 1880 and World War II, 80,000 birds were exported from nearby Dutch New Guinea in the feather trade. Spyer suggests that Indies birds also played a role in arguments between evolutionists and creationists in Europe: "if it was especially in naming that the religious and territorial projects of Europe initially tended to come together, from the eighteenth century on it was nature itself—revealing through study and knowledge the handiwork of God—that increasingly became the object of attention" (46). Could something so beautiful as a bird of paradise be merely earthly?

For better or worse, no bird as spectacular as the bird of paradise emerged from Togean Island treetops, but such a possibility helps explain Meyer's travels in search of new and unusual bird forms.[6] In Meyer's natural historical nature, there was no room for birds' usefulness or beauty to anyone but the scientist. The EuroAmerican collector was the only subject of natural history, and the formal properties of individual species its only object. Unlike their absence in Umbgrove's writings, Celebes people, as well, emerged through a specific logic in Meyer's work: the character of the peaceful flock could be found in its contrast to the warlike Alfurese village. The content here (peace versus war) is less important than the divide Meyer opens up between himself and those Celebes peoples he encounters. Different from Rumphius's curious nature, the human figure proposed through Meyer's natural history could not know their fauna in a scientific way and would only give a "pleasing answer" to the inquirer. The Celebes native emerged as incommensurable with the European scientist in Meyer's natural history—an unbridgeable rift, based upon "knowledge," had appeared.

In Umbgrove's Wake

Although 1990s biodiversity conservation would assume a different form than either Meyer's natural history, in which the Celebes person was unknowledgeable, or Rumphius's curiosities, in which Malay people contributed to how wonders were known, these earlier periods of nature-

making were important forerunners of the Togean conservation project. While Rumphius's ghostly shadow haunts later projects of species inventory, the late-colonial writings of Umbgrove and Meyer played a more material role in facilitating how the islands would be "discovered" as a site with valuable plants and animals. First, listen to Dr. Supriatna's description of why the Togean Islands were a likely location for late-colonial collecting: "Due to a violent history of piracy well into this century, the legacy of which we experienced on New Year's Day in 1996 when money was extorted from us practically at gun point, virtually no prior collection and study of fauna has been conducted on the central peninsula. . . . Early naturalists simply avoided the mainland while they collected fauna in the adjacent Togean and Banggai Islands" (Froehlich et al. 1998).

When, in the 1970s, marine conservation began to come into focus alongside more established terrestrial conservation projects, new sites needed to be identified. The fact that early naturalists had "simply avoided the mainland" in favor of the safer adjacent islands overdetermined these offshore sites as future locations for biodiversity conservation. It was Umbgrove's studies in particular that marked a trail directly to the Togean archipelago.

I learned about this history in Jakarta in 1997 when I interviewed one of the staff of the World Wide Fund for Nature, Nigel. Nigel recounted for me how the first marine conservation sites were found. "Where do you start in Indonesia?" he asked.

> We looked at the British Coast Pilots. We looked at maps and chose places based on types of habitat. Teluk Cenderawasih [in West Irian] was an example of this. Bunaken was identified through hearsay from divers. The Togeans were through Nick Polunin and Apriliani Sugiono. They followed Umbgrove's reports from the 1940s [sic]. Nick spoke Dutch and found old references to the Togeans. New spots were found through these old references and by looking at maps.

Traveling through Sulawesi before there were banks, roads, or phones was a challenge Nigel described to me as little changed from the early part of the century. With these practicalities, colonial natural history could not assist. Natural history also provided insufficient information upon which to found a modern conservation practice or a system of marine parks. In this sense, Nigel thought of Indonesia as a "big black hole":

> Imagine starting in Indonesia. You know absolutely zilch about the coastal waters in the biggest archipelagic nation in the world. Indonesia stretches over one eighth of the world and it was a big black hole. There were not even books, fish books, at the time. In '82 we had only one fish book and

studies of corals were in their infancy as well. Everything was centered in Java. There were no banks anywhere. When we would go out to do surveys, we carried millions of Rupiah. There were no boats anywhere, no transportation. The whole thing was a big adventure.

This sense of unbounded openness would change over time, however, since the practice of biodiversity conservation could not be created in just any way the scientists chose. Constraints existed in the form of prior conservation failures. Nigel problematized the early focus on biology; from the perspective of the 1990s, the exclusive attention to natural science in the 1970s and 1980s seemed misguided:

> We knew nothing about the marine environment. We would use hearsay and old reports. It was a case of trying to find the characteristics like protected species, intactness, centers of biodiversity, current threats. The whole focus was biodiversity. Now I look back with such embarrassment. We were so unrealistic. All we thought about were the science and environmental aspects. Everything was from a conservation perspective. There were no attempts to analyze resource use.

Out of these early explorations in search of "protected species, intactness, centers of biodiversity, current threats," et cetera, the relationship of Indonesia's people to biodiverse nature would increasingly form a problem for thought. By the 1990s the "wheel would turn" and attempts to "analyze resource use" would become as important to biodiversity conservation as searching out new species.

Ethnobiology

Hari and Laksmi began to study Togean people and their relationship to Togean nature as early as 1993. The work of the IFABS scientists to understand those peoples who lived in close proximity to the natures they wanted to save was indicative of a wider "social turn" in transnational biodiversity conservation. How they would approach Togean people was informed as much by the particularity of mid-1990s Indonesian national ideas of social and political difference, however, as it was by international norms for the scientific analysis of "society." Hari and Laksmi described their work for me on my own first trip to the Togean Islands in 1994, and I later read the reports of their study. Working in three Togean villages, including Susunang, they interviewed "mainly old fishermen," observed the species Togean people used, watched how these species were harvested from seas and forests, and fished and farmed with Togean people.

From the first, IFABS ethnobiology was directed toward solving particular problems that distinguished it from other instances of ethnobiology.

The scientists described their ethnobiology as,

> a study about the people's uses of biological resources that combines the disciplines of biological science and cultural anthropology. Its purpose is to discover examples of the potential of biological resources, the role of natural resources for people by recording biotic species, the useful technology of biological resources, along with researching the influence of government policy/wisdom in the management of conservation of biological resources. Through the lens of biology, ethnobiology can be seen as those uses of biological resources which are directed at the effort of improving conservation of natural resources. While from the lens of anthropology, ethnobiology can be seen as a system of knowledge which bridges patterns of culture and is aimed toward developing the people without damaging the potential of natural resources. (Yuliati et al. 1994:1)

Each instance of ethnobiological practice in the world must be understood for its specificities. IFABS ethnobiology had none of the emphasis on cognition usually found in American structuralist anthropology of the 1950s through the 1970s. Nor was it motivated by the search for new pharmaceutical compounds ethnobotanists were looking for in Mexico in the 1990s (Hayden 2003). In the Togean project, ethnobiology engaged scientific disciplinarity (merging biology and anthropology), it was aimed toward conservation of biological resources, it discovered things about Togean people (such as how they harvest resources), it was concerned with "development" of these same people, and it explored the influence of government policy on conservation. From this came a holistic multiplicity of discursive objects: ethnic identities, relations of school and work, settlement patterns and house structures, religion and belief, childhood, infrastructure, medicine, social organization, life rituals, knowledge and technology, plants and technologies for their use, et cetera.

One of the most important of these objects was Togean peoples' knowledge of plants and sea creatures and their uses. In their study, plants were divided into six types: vegetables, building material, ritual plants, cosmetics, craft plants, and medicinal plants. The numbers of each type were counted and listed, with the names given in several Togean Island languages: Togean, Bobonko, and Bajau (Sama). Notes were made concerning the uses of each plant: "the inner part of the tip of the plant is eaten as a vegetable"; "used for mattress filling"; "indication: liver, malaria"; "technology: root and stalk are boiled." Sea products were also listed and divided into nine types: demersal fish, pelagic fish, crabs and shrimps, turtles, shells, squids, sea cucumber, sea plants, and mammals. The study of sea creatures began with an extensive list of Latinate family names followed by Bajau, Togean, and "trade" names (Yuliati et al. 1994:41).

People were linked to nature through the category "religion and belief": articulations between "humans and nature bring to the surface religion and beliefs concerning the power of invisible nature" (15). Sama people were described as close to nature, and able to read its signs: "They enjoy viewing the blooms of beautiful sea flowers (*lalamei/nambo*) at the full moon. They look for currents, and raise nets, and if there are many durian falling in the forest, then many groupers will be in the sea." There were special prohibitions: "If a squid is blocking the way, it means that your earlier offerings were not received by the sea spirit. This means they have to return home so that they don't encounter a disaster." They possessed unique abilities: "Bajau people have the ability to dive to 30 meters depth and [for] several minutes without using tanks or compressors." And there were special knowledges: "Healing of sick people can be done by an elder that can see sicknesses (Mawi)" (16).

Formal and informal "social organization" was another category the scientists described. Formal social organization was "based on instruction from the local government." In its ideal form it "contains and coordinates the aspiration of the people." But it was not effective, they explained, because people are busy with their daily labors. Also missing was someone in the role of "motivator" to move things ahead. There are so many children, and very low education; there aren't any activities for youth, and marriage at a young age is normal. Informal organization included those organizations initiated by the people themselves: prayer groups, money saving groups (*arisan*), sports and music groups, and fishing and farming groups. Types of sports and children's games were listed: volleyball, soccer, ping pong, rock throwing, and kite flying (21).

How should we understand these inscriptions that bring to life the ethnobiology of Togean people? One possibility would be to route our interpretation through a critique of instrumental reason and the potential the ethnobiological data has to inform projects of governmentality. Nicholas Rose, for one, has described the practice of reflecting on characteristics of populations in European history, and the links between these reflections and governmental reason. He writes:

> In European thought from at least the eighteenth century, one can trace a variety of reflections on the special characteristics of discrete "nations" and "peoples," the possibility of writing the history of different peoples, anatomizing their differences, demonstrating how each participates in a shared tradition of customs, a shared descent, a shared language; how each has a set of habits, beliefs, mores, systems of law, morality and politics which partake in this common spirit. Over the course of the nineteenth century, a mutation occurred in this way of apprehending the collective existence of a people. Nations were now seen as populations of individuals with particular charac-

> teristics, integrated through a certain moral order. But more significantly, this way of understanding the subjects of rule as subjects of morality was linked to a plethora of interventions into the economy, the family, the private firm and the conduct of the individual person which sought to shape them in beneficial ways whilst safeguarding their autonomy. (Rose 1999:101)

IFABS social science likewise linked biodiversity conservation to a moral order, and this order rested similarly on the characteristics of a Togean population. Hari and Laksmi and the other scientists wanted to make biodiverse nature comprehensible and desirable for Togean people such that they would adopt an ethical stance (preserving biodiversity) in relation to Togean nature. To do this, Togean people themselves would be analyzed, studied, classified, described, and ultimately formed as an object upon which conservation and state governmentality could act. In the process they would become "indigenous people."

From this analytic perspective, I find the IFABS ethnobiological project more reminiscent of Meyer than Rumphius. Rather than opening their nature-making project to other possible natures, the scientists' polyglot species lists incarcerated permissible natures within forms recognizable to natural science. Lists of nature's things began with Latinate terms, and Togean people could only contribute nouns within an already established structure of representation. Fishing and farming technologies were opened up to moral evaluation, and these techniques adhered to specific ethnic groups. Once the social organization of the group was defined, Togean people were diagnosable according to bourgeois norms of social development. Sama and Bobongko people were found wanting in terms of social, religious, and medical organization, and in need of intervention by conservationists necessarily aligned with state norms of citizenship. Like Meyer's inventory of birds, these research findings forced a civilizational divide between scientists and Togean people.

Other aspects of the scientists' descriptions of Togean people remind me more of Rumphius than of Meyer, however. Different from the interpretation of "local people" witnessed in many other international conservation projects, Togean people *had* knowledge at all. Since local people were often presumed to harm nature out of ignorance, to come down on the side of the existence of Togean expertise was controversial to begin with. As Dr. Supriatna once said, "education in these Islands is low, low, low. Gunung Halimun [in Java] has access to education, access to television, and access to newspapers. They have information from the outside world. But here people can be misled. There is miscommunication many times." In this context, IFABS' ethnobiological work was at the minimum a statement in support of Togean people.

While we might question the logic of an advocacy that depends upon the same social distinctions that exist within the theories of reason that indigenous knowledge advocates wish to subvert (Agrawal 1995), most of what Togean people knew or did was not described as "destructive" by Laksmi or Hari; blast and cyanide fishing, for example, actually receive very short mention at the end of their text on ethnobiology. Additionally, Togean peoples' engagements with nature were viewed as not merely economic and instrumental: "they enjoy viewing the blooms of beautiful sea flowers (*lalamei/nambo*) at the full moon." As in Rumphius's curious nature, the scientists described people by name and related their stories: "The family of Mbo Janiba left to go fishing at 8:00 at night and returned at 11:00 am after they felt they had caught enough *trepang* [sea cucumber]" (Yuliati et. al. 1994:19). And, when Togean people did not respond correctly to government programs, they did so for a clear and positive reason—they were too industrious to have time for such spurious activity (21).

Through their ethnobiological work, the scientists' interpretations had begun to diverge from the logics of transnational conservation biology with its threatening human that puts nature at risk.

Participatory Spatial Planning

In May of 1997 I joined a planning meeting at the CI office in Jakarta, where the Togean Island park was discussed in new spatial terms. Conserving biological diversity involves the reorganization of space through such tactics as gazetting territory for parks, and creating use, non-use, and buffer zones. Some spatial strategies that isolate flora and fauna from human use, or connect wildlife across corridors, respond to the spatial needs of plants and animals themselves.[7] As Dr. Supriatna had explained, "big charismatic animals have a wide home range. For example, the elephant has a 90-square-kilometer home range. If you're protecting them, you're protecting a lot of small plants and animals." Others, such as the creation of special areas for people's engagement with natural features, attempt to ameliorate the threat humans pose to nature by limiting how people interact with their surrounding environment.[8]

At the meeting that day were Hari, Laksmi, Yakup, and Budi from IFABS, and a representative from CI, Washington, D.C. New and different indigenous knowledges were to be sought, and the uses to which these knowledges were to be put were transformed. Togean peoples' ideas of space and place were the form indigenous knowledge would now take. The scientists described the new configuration this way:

One site of the spatial plan: The waters surrounding Susunang Village. © Karl Taylor (*www.karltaylor.co.uk*).

> Indigenous knowledge has gained substantial attention since [the] last decade. After it was neglected for decades, a movement to revitalize the knowledge is growing as a way to empower local and indigenous communities and to improve or complement scientific knowledge including those in natural resources management. Indigenous knowledge on natural resources or ethnoecology has evolved for decades or centuries as part of indigenous cultures and much of it has been proven to be sustainable. The knowledge includes how indigenous communities use space (land and sea) with its philosophies behind it. (IFABS 1997:1)

The question would be how to involve Togean people in the planning process by incorporating "indigenous spatial knowledge" into the plan. The scientists wrote,

> Sustainable use and conservation of natural resources is a major issue in the last few years. Prior to the implementation of those activities, spatial planning is a crucial stage to ensure sustainability and justice. Unfortunately, in most cases, especially in developing countries, spatial planning is carried out merely by [the] government with inputs from scientists. Local communities who are the most affected part of the spatial plan, often in negative ways, are usually excluded in the planning process. Therefore, ways should be created to include local communities' interests in the plan. (IFABS 1997)

This articulation of indigenous knowledge, sustainability, and social justice brought into focus new questions about scientific management.

Working on a whiteboard, we came up with a series of "threats," "programs," and "priorities" for future Togean work, and we discussed how to involve Togean people and the Sulawesi government as "stakeholders" in the spatial planning process. Our conversation began with the difficult problem of participation:

HARI: Traditionally the government has been planning from the top down and neglecting *hak ulayat* [traditional rights]. There has been no community involvement.

LAKSMI: But there is a tension between our role and the government because our work is trying to get the government to do something it doesn't normally do.

YAKUP: Yes, the government can be suspicious of us working with the community.

HARI: We have the problem of the government thinking we are working too closely with the people, but also of the people thinking we are working too closely with the government.

The problem of spatial planning would be to balance the interests of the government with the interests of the people. Many of us who studied Indonesian politics during the Suharto period imagined there was little room to maneuver within Indonesian state mandates. But elite Indonesians, like the IFABS scientists, understood the state as a flexible institution, and had a better sense of the possibilities of working within its confines. The solution the scientists were developing necessarily involved assisting the state from within (rather than confronting or opposing it directly) while hoping to change it in the process.

Several changes in the world of Indonesian conservation NGOs and in state-society relations had come together at once to make possible the idea of spatial planning. First, a series of emerging laws, beginning with the Spatial Planning Act of 1992, was issued by the Government of Indonesia (GOI 1992). By 1997, the national legislature (*Dewan Perwakilan Rakyat*, DPR) had developed these laws to give communities an opportunity to contribute to regional spatial plans. Second, CI and the other big conservation organizations had begun to encourage mapping projects that used Geographic Information Systems (GIS) as a tool for conservation planning (Harwell 2000). Third, a discourse of "counter-mapping," designed to use maps to argue on behalf of community rights, had recently emerged within Indonesian NGO networks (Peluso 1995). Additionally, conservationists had discovered an extensive system of property rights and temporal and spatial resource exclusions in the Maluku Islands called *sasi*, and many people were looking to find such a thing in other places (Zerner 1994). Finally, new possibilities for political advocacy were begin-

ning to open in the last years of the Suharto era. These possibilities were reflected in our conversation that day:

> HARI: Policy in Indonesia is not like black and white. It depends on how you present it. There is always leeway. The challenge is enrolling local government as an ally.

> YAKUP: Our consortium has a facilitating role, a role providing biological information, and it is also a stakeholder in what happens. The information Bappeda [*Badan Perencanaan Pembangunan Daerah*, the Regional Development Planning Board] has is very basic, that's why this group has a role. The consortium will provide technical assistance to Bappeda. The result will be a map and the introduction of GIS as a tool.

A dream of seamless translation between indigenous knowledge and state management plans guided our conversation. Togean communities were imagined as homogeneous and autonomous, and "participation" could be turned into a problem of technique: "Documentation of [the] community's natural resource management and spatial use knowledge . . . will be carried out by encouraging local communities to document their knowledge (written and orally) while project staff will only act as facilitators" (IFABS 1997:3). This imagined community would be naturally interested in the project and would grant approval for it: "With [the] community's approval, all maps and other information will be digitized using GIS." And the community would be capable of communicating its message directly to local government: "Production of [a] community version of [the] spatial plan based on maps and other information resulted from previous works [that the] community can exercise to produce their version of [a] spatial plan. They can bring results from the exercise to [the] DPRD (local parliament) and government as a base for further discussion and negotiation" (IFABS 1997:4).

Laksmi acknowledged the problems inherent in imagining a harmonious community that could provide unitary spatial knowledge, or a community member who would willingly desire to contribute to a newly managed and bureaucratized landscape, and who was able to argue with the state using scientific methods.[9] She explained, "the word *partisipasi* changes between the government and NGOs. They aren't meaning the same things."

Given the ambivalence many scientists had to Togean peoples' occupation of the proposed conservation area, however, and considering the difficulty of directly confronting the government on the basis of political or citizenship rights, there was a remarkable transformation in the move from ethnobiology to spatial planning. The Togean Island person was

now configured as an Indonesian citizen with the right to participate in the political process of spatial design. Moreover, this person would exercise discretion, judgment, and thought. Despite some bumps in the road, the wheel was beginning to turn from "indigenous knowledge" to a sense of justice and care for Togean others, and the Togean citizen would now contribute to thinking through and finding solutions to the problem of environmental change. The scientists argued,

> Within the last twenty years big commercial companies have been active in the archipelago and created conflicts with local communities. Those companies exploit terrestrial and marine resources often within areas that have traditionally been extractive areas for the communities. And recently, as tourism is growing, some companies are building tourist resorts. New ventures are planning their investments and are applying for plots of lands or sea for their activities. These are growing rapidly while a detailed spatial plan is still not available. If such a plan cannot be produced soon, greater conflicts will occur which will harm local communities as they are more vulnerable in such conflicts. (IFABS 1997:1)

In nascent form within the spatial planning project, the Togean person had been refigured from the owner of knowledge in the form of "words for things" into a rights-bearing Indonesian citizen with the capacity for reason and thought. This move captures the aura of Rumphius, who, while working for the VOC, was simultaneously one of its most outspoken critics.[10]

Friendship and the Limits of "Knowledge"

Laksmi once told me a story of how, when she was still a student at the University of Indonesia studying for a degree in biology, she conducted an experiment in Bali on seaweed. She was looking into the possibility of growing seaweed in brackish water ponds called *tambak*. Tambak are also used as shrimp ponds, and the existing network of ponds presented the possibility of growing shrimp and seaweed together. Seaweed (*agar agar*) is used in making deserts because it gels without refrigeration, and it is also exported for use in cosmetics and ice cream. Laksmi's experiment was designed to address the problem of seaweed biology and sustainable aquaculture. But seaweed has encountered many difficulties in Indonesia, including disease and inadequate rates of growth (Sievanen et al. 2005). Laksmi herself was having trouble getting it to take hold in her field trial; crabs were eating the seaweed stock before it could mature.

Laksmi tried everything to keep the crabs away, and she was worried about not getting a result she could use in her thesis. She began to discuss

the problem of crabs with a Balinese farmer she had befriended who lived near the tambak. The farmer told her that the seaweed would not grow because she hadn't asked permission to build her tambak from the land's "owner" (*tuan*). By this he did not mean its human proprietor, he meant the spirit that inhabited that place. He helped Laksmi to make the appropriate offerings of "something like a chicken and some ceremonial rice" when, low and behold, the crabs disappeared, and Laksmi's seaweed began to thrive.

Laksmi enjoyed telling me the story of how a ghost helped her seaweed research. What she did not claim in her telling was that she was helped by Balinese "indigenous knowledge." In describing her experiment it was not necessary for Laksmi to produce something called "Balinese culture," nor did she find it useful to turn offerings into a form of scientific data. In fact, she could not quite recall the propitiatory recipe they had used . . . just something about a chicken. The existence of the tambak spirit did not threaten to open a hemispheric divide between the modern and the primitive, the rational and the spiritual, or to incarcerate Balinese within the realm of the local. For whatever reason, Laksmi did not feel the need to set up an opposition within which she would stand on the side of universal reason and intellect while her Balinese neighbor would inhabit the space of local knowledge or mystical belief.[11]

In much of the literature on what has variously been termed "indigenous knowledge," "indigenous technical knowledge," "traditional ecological knowledge," "local knowledge," et cetera, the "indigenous" and their "knowledges" have appeared as given objects in the world amenable to a priori definition. Indigenous knowledge has been described as local, orally transmitted, practical, empirical, shared, and experiential (Banuri and Apfell-Margalin 1993; Ellen et al. 2000). Such definitions assume a fundamental opposition between plural indigenous knowledges, and a singularity called "Western," "modern," or "global" scientific knowledge. Those who theorize the field frequently subsume indigenous knowledge under scientific knowledge by certifying that indigenous knowledge is scientifically valid. In the process, indigenous knowledge has become a form of instrumental reason aiding in projects ranging from economic development to conserving biodiversity. It is hard to refute the importance of this liberal move in a world where technocratic solutions and governmental rationality devalue forms of thought not recognized as either scientific or managerial.

The proposition of a radical disjuncture between the indigenous and the scientific can be contested, however. For example, Richard Grove has observed how the system of botanical classifications used in Cochin in southwest India traveled to the Leiden botanical garden (from which Linneaus took much of his material) through the *Hortus Botanicus* of H. A.

van Reede. Grove calls the *Hortus* a "profoundly indigenous text," and claims print technology allowed South Asian classifications to travel to Europe along pre-existing, privileged European botanical networks to become what we now think of as "Linnean" taxonomy (Grove 1995:78). For contemporary India, Akhil Gupta describes the hybridity of scientific and indigenous knowledge in Alipur, a village in northern India. While Alipur farmers use chemical fertilizers, electrically driven tube wells, and bioscientifically developed hybrid seeds, they also engage in discourses and practices of "humoral" agronomy (Gupta 1998:155). In breaking down the separation between the indigenous and the scientific, "indigenous knowledge" is revealed to be a system for classifying *peoples* rather than what they know.

In Togean ethnobiology and spatial planning, we can see the shadow of Rumphius hanging over Laksmi and Hari's desire to heal the divide between scientists and Togean people. The rift that first opens in Meyer's descriptions of peaceful birds and warlike Alfurese could not be easily erased in the conservation project, however, since neither Togean people nor Indonesian scientists were readily able to disrupt the international norms and standards of biodiverse nature with its threatening human that puts nature at risk. The very act of creating indigenous knowledge, subject simultaneously to evaluation by scientific knowledge, has meant that the "scientific" will never be fully unsettled by the "indigenous." Incommensurability was one consequence of the social turn in transnational conservation science.

But we can also see in the scientists' work a utopian vision linking reason and the nation with the possibility for citizenship and inclusion. In employing ethnobiology or spatial planning, not simply as instrumental reason, but as an expression of loyalty to Togean others as members of the Indonesian polity, IFABS scientists demonstrated their faith in the nation-form as site of freedom (cf: Cheah 2003). It does not serve us well to see the indigenous figure of Indonesian science as precisely identical to that "Other" formed in dialogue with European Romanticism or northern scientific rationalism. Just as Rumphius aligned himself with Malay thought on Ambonese nature, Indonesian biologists manipulated indigenous knowledge to form a new political imaginary and an object of hope.

I too would like to hold at bay the space of indigenous otherness. I want to argue from the position of an anthropologist and science studies scholar that we should think of Laksmi's solution to the crab problem within the ambit of normal science. But the ordinariness of science's ability to coexist with ghosts, or the nonobjectification of knowledge called "indigenous," "local," or "ethno," depends on something more than science. "Friendship" might be one solution to the problem space of the hemispheric divide. Such friendship was often an unexpected, unintended,

and unremarked outcome of Indonesians' science in the Togean Islands. Friendship creates a new geography that Paul Rabinow refers to as the "*philia* site." "I propose that a primary site of thinking is friendship (philia)," he writes. "Such a formation sounds strange, as friendship is a relationship and not a physical place. Today, we have lecture rooms and conference rooms and meeting rooms and classrooms and offices and studies; the Romans had rooms for friends" (Rabinow 1996b:13).

Perhaps Rumphius also had such a room in Ambon where he came together with Patti Cuhu to discuss Indies flora. Rumphius wrote of his friend, "this wood was shown to me for the first time . . . by Patti Cuhu, a Regent or Orangkaya of the Hitu village called Ely, a Man experienced in the knowledge of plants, who has helped me a great deal in this work, wherefore I deem it only decent, to commemorate him here" (Beekman 1999:ciii). Similarly, Yakup reminded me of the pleasures we shared with Abo, Musir, Suala, Narto, and others in Susunang while gliding through warm waters in search of sea cucumber together:

> *Celia,*
> *How are things? . . . Last February I had a chance to travel again to several Togean villages including Kabalutan [Susunang] with CI. Out of the work, it's so so nice having even a very short time in there. Many* Puah *and* Mbo *[parents and grandparents] asked about you. They miss you so much. I had a very great great time (only two nights) in Kabalutan and went one night* balobe *[sea cucumber collecting] with Abo, Musir, Suala, Narto, and some others. That's all the news from Kabalutan for now. Sukses. . . ,*
> *Yakup*

How productive would it be to think of *holothurians* from the "philia-site" rather than the site of indigenous knowledge? Might this not be an excellent location from which to begin to think about the diversity of sea cucumber life forms, or to overcome the instrumental imperatives of species driven biodiversity conservation? For the preservation of the diversity of life in all its forms, we may find that sharing thought is superior to collecting knowledge, or that forms of friendship have advantages over forms of reason.

PART TWO

Togean Cosmopolitics

THROUGH THE IDEA of global nature-at-risk, biodiversity entails a cosmopolitan imagining of community beyond the nation or region in which any given instance of species uniqueness or diversity is found. For Indonesian scientists deeply committed to the idea of the nation, rather than subverting the idea of "Indonesia," the practice of biodiversity conservation reinforced a project of Indonesia-in-the-making. And despite these transnational or national ambitions, biologists developed the case for threatened global nature through the regional microsite of the Togean Islands. In the contingent assemblage of plants, animals, and people that came together in the Togean project to produce a claim for universal nature, the relationship between "nature" and "the human" was put into play. The generalized humanity for whom nature was being saved could only be developed, however, by way of contrast with the figure of the specific human who put nature at risk.

And what of Togean people who were the object of this attention? What were the natures, knowledges, ideas, or problems at stake for them? In part 2, *Togean Cosmopolitics*, I turn to questions of knowledge, nature, reason, and the human for Sama people who live in the Togean Islands. In a sense, this book begins again. Rather than starting from the universal (biodiversity) and demonstrating its ability to produce the national (Indonesian science) and the local (endemic Togean species and Togean people), we will start from the local (Sama people and their natures) and work our way out, via the nation, to an encompassing universe (Sama cosmopolitanism). If not through transnational science, how do Sama people come to fashion a world around them and in relation to what transregional connections and constraints? In part 2, I (re)present the Togean Islands, not through any claim to greater accuracy or truth, but in an attempt to bring into focus the relations of force, meaning, and scale within which natures emerged for one particular group of Togean people, the Sama residents of Susunang village.

The themes of part 2 are foreshadowed in a story sung for me by Mbo Biba one night in the Togean Island hamlet of Kilat. I first met Mbo Biba when I was on a *pongkat* (S) trip with Puah Umar in search of the rare and valuable white sea cucumber. Mbo Biba himself was on a *pongkat* catching and curing salt fish. Umar and I had anchored in a cove well

known as a hideout for pirates, and Mbo Biba had observed us for a long time before he approached us in our *leppa* (S). When he came by, he shared some herbal medicine (*jamu*) he kept in an old bottle to keep him strong while he traveled. It was a fortuitous meeting. Mbo Biba was known as the last Sama person in the Togean Islands to be able to recite *iko iko* (S), a type of song poem about the past, and I had known of him for a long time. The last time Umar had heard *iko iko*, he told me, was before he was circumcised.

We arranged to meet at Umar's mother's house. As Mbo Biba started to chant out his stories of "Si Bawang" and "The Girl in the Sea Foam," everyone present was affected by their beauty and many began to cry. When later Umar decided to listen again to Si Bawang, tears came down his face as he sat under the headphones of my tape recorder. Through his narrative, Mbo Biba connected Sama people to plants and animals, to the former Bugis-Makassar Sultanate of Gowa in South Sulawesi, to ineffectual medics and midwives provided by the Indonesian state, to Dutch colonialism, and to the cosmopolitan world where, he speculates, "even in America, even in Russia, everywhere they are found by the sea, our Sama houses." This is Mbo Biba's story *Ningkinda ma Buburah* (The Girl in the Sea Foam):

Si tempo nə, a ma rompong, tikkə ma tanah Boges ka dilau. Ketummu nə leppa rompong iru buburah bigi rumah ma dilau.

One time, at the fishing float, coming from Bugis land heading out to the sea. Next to a fishing float, was found some sea foam as big as a house.

Nia ningkində missa dadarua melssou nə ma di alang buburah. Mole iyyə ka Datu ma Boges.

Inside the sea foam was a girl, and there was no one more beautiful than she. He [the one who found her] went home to the Bugis King.

Tuntu'pang nə Datu ma Boges, iyyo' nə, "nia ki'tə' kami buburah missa darua basar nə, melassou nə, nia ningkinde ma dialang."

He spoke with the Bugis King, he said, "there is something we saw, sea foam of which there is none larger or more beautiful, and inside it there is a girl."

"Dadi," iyyo' nə, "lamu missa ningkində iru, dibono kang." Sudah!

"Well," the Bugis King said, "if the girl isn't there [when I look], you will all be killed." There!

Pore iyyə palimbə, ma ka Datu

He returned [to the sea], the

ne. Naginda iyyə tikkə ka di laut iru, nia masih.

one who had come to see the King. He looked for her, he came to the sea [again] and she was still there.

Ah! mandiru ndah dibowə ne kunyi, buas ma dikunyitang. Diamburang, nggei lagi iyyə lai.

Ah! there was brought yellowing and rice to be yellowed [To be tossed over her in welcoming]. It was tossed over her and she didn't run away.

Dadi patuku iyyə. Alla' nə tanggan nə batiru ne, "paitu ko dutai ka leppə." "Aku," iyyo' nə, "itu Datu."

Then he [the king] approached her. He grasped her hand and said "come here aboard my boat." "I," he said, "am a king."

Nggei babaong.

She didn't say a word.

Tikkə mandiru alla' nə, padutai nə, dibowə ne ka darə'. Dibowə ka darə' dipato' ne banderə ula ula

From there he took her on board and carried her to land. She was brought to the land and the *ula ula* [the Sama banner] was raised.

Takudə' ne memong ne manusia ma darə', ma Tanah Boges. Iyyo' nə, "nia ne, beleh, dibowə paitu." Tikkə mandiru dipadutai ne ka rumə, nggei babaong.

All the people from the Bugis land were surprised. He said, "she was there [at sea in the foam] and I brought her here." From there [the shore] she was brought up to the house. She didn't speak.

Bittə bittə ne bonə pandan nə ale Datu Boges. Pandan nə ale Datu Boges. Sudah ne, dipanikkə ne itu Datu. Dipannikə Datu itu nggei babaong.

After a while she was married by the Bugis King. Married by the Bugis King. Even after she was married by the King she didn't say a word.

Bittah ne. Ngidang ne. Ngidang ne iyyə. Noro ne iyyə ka dilau. Noro ka dilau.

And she became pregnant. Pregnant she became. She pointed to the sea. Pointed out to sea.

Sukar memong ne, ai ko itu ditoro ma dilau?

She made it difficult for everyone. What was it she was pointing to at sea?

Na pore a'ə ka dilau ngala kimə, batunang, bikah nə anu

Everyone went to the water to harvest giant clams and *batunang*

idang idamang ma dilau. Na dipu'po' ne paitu.

[*Labidodemas seperianum*, like a sea cucumber] so she could satisfy her cravings from the sea. The food was gathered together.

Kimə iru je ma diala', nginta nə, sampei basar, sampei ngana'. Nggei babaong.

Only giant clams she took and ate, until she was large, until she gave birth. Still she didn't speak.

Ngana' iyyə mandiru. Tapaluang ko itu, missa daulu, missa kang.

She gave birth there. Out it came right there, and [speaking directly to his audience] there wasn't [yet] our old folks, there wasn't [yet] you all.

Noro umpə ka bullu iyyə. Pore ka bullu iru dialla' bikah nə kuli' kayu. Dibowə paitu sapala nə bagu. Pugei nə batitu tanggan nə. Terpaksə nadiki'tə' itu, ponso itu.

She pointed to the mountain. So they went to the mountain and brought back some bark. They brought back here some oakum. She rolled it in her hands [Mbo rolled his fingers as though he was rolling a cigarette]. They cut it like that, the umbilical cord.

Dadi, kitə bo tikkə ma taung ka taung indah bagu itu ne jə. Bobonə ne nggei lagi melalui mantri. Nggei lagi iyyə. Ah, batiru.

Thus, then, from year to year we [Sama] look for only oakum. We don't use the government nurses, not them. Ah, it's like that.

Nangis ana'ana' itu puli. Nggei babaong. Nangis ana'ana' itu puli, noro ka dilau.

The baby cried continuously. She [his mother] didn't speak. The baby cried on and on, and she pointed to the sea.

O'. Sukar memong manusia. Dipaksə dipugeiang ma sidi boe rumə Dipugeiang, kang, Datu itu.

Oh, it was difficult for everyone. They were forced to build a house at the edge of the sea. So, you all [the audience], the King made it.

Ah! Nggei lagi nangis ana' ana'. Ah!, mandiru uyə uyə iyyə.

Ah! The child didn't cry any longer. Ah! There [by the sea] she [the mother] sang and sang.

Iyyo' nə, "kau," iyyo' nə,

She [his mother] said, "you,"

"arung nu batuah nə ma Bone 'Somba di Gowa'."

she said, "your name in Bone means 'Somba from Gowa.' "

Mau, iyyo' nə, kami ndah Datu jə du. Mau, iyyo' nə, aku ndah Datu jə.

Even, she said, we are certainly royalty too. Even, she said, I am certainly a queen.

Sudah basar basar ne ana'ana itu. Matei ne Ua itu, basar ne ningkilla nə. "Ah!" iyyo' ana' nə ma darə', iyyo' nə, "aku bəkə nanangah kerajaang."

Then the child was grown. His father was dead by the time the son was grown. "Ah!" said the child on the land, he said, "later I will hold the kingship."

Iyyo' ore umpə ma dilau. "Nggei!" iyyo' nə, "bo pager lillə."

Those from the sea said to him. "No!," they said, "we are all men together."

"Nggei!" iyyo' nə.

"No!," he said.

Terpaksə bono, nabono. Terpaksə nabono, dutai boe, tikkə ma dilau. Dutai boe rumangi bembe ai.

They were forced to war. As they were forced to war the water arose from the sea. It rose so high that all the goats were swimming.

Dipato banderə panyoroh, bandera potei dipato'. Ah! Nggei lagi dadi bono.

The flag of surrender was flown, the white flag was flown. Ah! There was no longer a war.

Dadi samə samə ne marintah. Iru ne bonə kitə daulu. Sama itu ne nggei mina dijajah ale Belanda.

So together we govern. That was us in the time before. Sama people were never ruled by the Dutch.

Dadi memang benar benar itu. Mbo Mbo tə tatumu ma dialang buburah kitə Sama itu.

Thus, this story is certainly a true one. Your ancestors were found in the sea foam, these Sama.

Cobə, kono, tele tə tele tə maninggə la'hə' nia ke kitə Sama itu bonə simatə ma siddi boe, ruma tə.

Try, you all [who are listening], to see wherever there are villages, isn't it true, we Sama can find each other at the edge of the water, and your houses [too].

Missa Sama itu bonə ore ma roma, lalangkar tadakau da musih ma siddi boe.

There are no Sama over there in the forest and rarely one or two are found away from the sea's edge.

Mau ma Amerika, mau ma Ruslang, mau madingga tetap masih di boe, ruma nə Sama.

Even in America, even in Russia, everywhere they are still by the sea, Sama people's houses.

Chapter Three

EXTRATERRESTRIAL OTHERS

Of the statement that they are afraid of death when they come to support themselves on land, the opposite appears on Tabungku, since one finds there a whole [Sama] village where they are thriving well on a mound by the sea, so that I believe that if one were to investigate, one would find more of such villages, and that they all can thrive as well on land as on the sea.

—Robertus Padt-Brugge, Regent of Ternate, 1676 [Sopher 1977]

SAMA PEOPLE, in 1990s Indonesia, were imagined as extraterrestrial others: both living beyond the land and alien. They were called Indonesia's "floating peoples" (*suku terapung*) and were seen as one of many "alien ethnicities" (*suku terasing*) scattered across Indonesia's far-flung hinterlands. Representations of Sama identity emerged within the ambit of two ethnographic facts. First, next to the Chinese in Southeast Asia, Sama belong to the most territorially expansive ethnic group in the region, living as they do along the coasts and strands of much of insular Southeast Asia. Unlike Bugis or Mandar (other Southeast Asian "seafarers") who have well-defined homelands, Sama communities are found throughout eastern Indonesia, Sabah, Malaysia, and the southern Philippines.[1] Second, in addition to the unusual feat of being all over the place, Sama people are known for a particular lifestyle that seems exotic to nearly everyone. Namely, they occasionally live aboard small canoes in which they travel looking for sea cucumber and other marine products. From this, many have assumed that land is unimportant to Sama lifestyles.

Togean Sama people *do* sometimes make periodic extended collecting trips around the Togean Islands living aboard their canoes (*leppa*[S]) or living in shelters (*sabua*[S]) over coral reefs for several days or weeks at a time. During these trips, called "*pongkat*," people collect marine products, largely sea cucumber and fish, for sale to Bugis shopkeepers and mobile traders. Living aboard a boat is never a sign that one lacks connections to a village or to the land, however, or that one is drifting aimlessly or endlessly in some Robinson Crusoe adventure. The pongkat is a voyage out and back, as it were, from and to home. Moreover, not all Sama people go on pongkat; many prefer to spend their time in the village or in their gardens.

In writing about Indonesian scientists, I have avoided over-determining their relationship to Togean nature by assuming we do not know the iden-

tities of scientist or science, culture or nature, species or genus, before we commence, and we can track relations between Sama identities and Sama natures in a similar methodological fashion. To do so would be to think of Sama peoples' mobility aboard boats as productive of both identity and marinescapes without naturalizing Sama identity as "sea nomads." What follows is a description of a particular pongkat that I made with Puah Umar, my friend and fishing partner in Susunang, for two weeks in April of 1997.

My day with Umar began several hours after sunset, rowing out to the reef in the quiet of the night. There were fourteen boats altogether on this pongkat. The only sounds were the fishers' conversations and an orchestra of paddling as paddles dipped into the water and drummed against the sides of our boats. Floating across the night sea was like slipping across the surface of a black and white photograph; there was no tint of color to the surrounding marinescape. The sky was filled with stars from horizon to horizon and the moon was absent. It was *bulang duƏmpulu pitu* (S), the twenty-seventh night of the lunar calendar, the phase of the moon when certain sea cucumbers "stand up" on the sea floor.

Approaching our destination, White Point Reef (*Sapa Toro ma Potei* [S]), we bumped across a shallow, hitting the bottom several times and having to back up and go around the obstructions. Once there, Umar stood on a coral head and pumped up the hurricane lantern that he had hung from the bow of the boat. Billowing flames rose up from the front of each boat around us as lamp burners were primed. Suddenly there was color: first yellow flames, and then the sea bottom lit up in a dull blue green. Below the surface were surreal green and yellow shapes of branching fingers of coral and smooth sand patches. The stars seemed to go out as the lanterns came on, and only the few planets were left visible overhead.

I slid over the edge of the canoe and into the water. The water felt warm to me, and I learned that although I would say that air temperature changes from night to day, Umar described the sea water as changing from warm at night to cold during the day. After an hour or so, I would always start to get cold in the water. Umar told me that to withstand the cold I should eat bitter things, and he mentioned the names of two wild forest tubers, *kuntang* (S) and *bantumi* (S), that are good for building cold tolerance.

As I looked across the water's surface, I saw heads beginning to bob, first looking down at the bottom and then coming up for breath. I could hear the sounds of voices bubbling at the water's surface as people called back and forth so as not to lose each other in the dark. They called out the names of kinds of sea cucumbers as they caught them: *Rege*! (S), which

The circular lume, by Celia Lowe.

means the thorny one, or *Haji*! (S), which means the fat one with the white spot looking like the hat of a Muslim man bent over in prayer.

Swimming inside the circular lume of the lantern's green glow where it had forced back walls of dark sea water, Umar and I moved along, with him tugging our heavy boat behind us. Looking sideways across the water column I saw muscular bodies descending to the bottom in halos of translucent green light. When I dove, I went headfirst but Umar and the other divers would push the water upward with their hands and fully submerge before turning turtle and heading for the bottom. I kicked my legs together in unison, whereas they had learned to kick with one leg at a time in syncopation. Periodically, Umar or I would spot a sea cucumber, and then one of us would dive down to the perimeter of the lantern's shallow brightness to pick it up with our fingers. We would also see other creatures when we dove. I brushed past a lionfish with its dangerous spines extended to paralyze its prey. Umar scratched his head for good luck to keep a large squid from fleeing and then he caught it with his spear gun.

All together, Sama fishers harvest about twenty different kinds of sea cucumbers from Togean waters. Some sea cucumbers, when caught, throw out their guts in sticky streams. *Gamma batu* (S) has a flesh that easily deteriorates in contact with other objects. *Bale potei* (S) and *Nanas* (S) are very valuable, while most, like *Lolosong* (S) or *Jejeper* (S), are not. Sea cucumbers crawl out from under coral rocks and ledges at night onto sand and seagrass beds, where they feed off the organic debris on sand particles. Every month during the pitch black of the twenty-seventh

moon, one particular species, Gamma batu, is especially visible and abundant. It "stands up" to spawn with its sucker feet attached to a rock at one end while the other end sways erotically in the current.

At the end of several hours of collecting, it was the middle of the night. We boarded the boat, washed off with a liter of brackish water, and paddled to a sheltered spot close to an isolated beach. On the way there hundreds of tiny minnows jumped out in front of our bow in random form. Narto, in the boat next to ours, said that the fish were attracted to the lanterns, but it seemed to me as though we were rowing directly into a chaos that was there before we were. Umar left his lantern on to warm himself and smoked a cigarette. We heard a Gar (*Sori*[S]) skipping across the water. Long fish with sharp beaks and teeth, Gar jump across the surface and will pierce your body if they hit you. Narto still had teeth working their way out of his knee from an encounter with a Gar five years earlier.

At the anchorage I pulled off my wet shirt and bicycle shorts inside my sarong and rolled out my sleeping mat. Luckily, our leppa was just long enough for Puah Umar and I to sleep toe to toe without embarrassment. There was only time for a short sleep before daylight came. At first light, all the boats were pulled up on shore, and the long, hot process of curing the catch commenced. The beach was a scrappy little patch of sand covered with coral rubble, but it was a place no one claimed. Umar gathered dead mangrove limbs for firewood and he built a platform out of sticks to hold a big aluminum wash basin of water over the fire. Then he boiled the sea cucumber. The air was hot and unpleasant on land under the equatorial sun with fires and smoke all around us.

I studied the different types of sea cucumber the fishers had caught and the special preparation each required. Gamma batu had their intestines removed before cooking, and the guts came out in a white and very sticky mass on the ground. Puah Padi held his Gamma up after slitting the end with a knife to elongate the animal; the longer ones are more valuable, he explained. The skin of another sea cucumber, *Bale potei* (S), was scraped first to rid it of sharp nubs. Bale potei is a rare and expensive sea cucumber and exciting to find. Another kind, *Karidau* (S), had a bracket of sticks inserted into it to help it keep its shape. Most sea cucumbers, however, were simply washed before and after boiling. Then the sea cucumbers were put over the fire to be smoked. While they were smoking the sea cucumbers sizzled, and Umar joked, "O, the sea cucumbers are crying" (*O, nangis bale* [S]).

Finally, we set the sea cucumbers in the sun to dry for the rest of the day, and we cooked a meal. Umar told me you shouldn't grill shellfish, like lobster or crab, on the beach when you are away from the village. If you do, this will bring the land and mountain spirits down who are

attracted to the smell. They will make their presence known by making you sick. We ate sago, which is good to take on a pongkat because, unlike rice, it doesn't require water to cook. I had brought vegetables to eat, but that was unusual. In the afternoon, the hottest part on the day, we anchored again and slept in Umar's leppa which was cool and shady inside. That afternoon, I wrote journal notes and drew a picture of the inside of our boat.

In the early evening, Umar told stories in a loud voice so that everyone in the anchorage could hear. One was about tourists. Once, near Katupat, some tourists had decided to spend the night in a cave. He asked them, "Aren't you afraid to sleep here? There are ghosts." They said no, but in the middle of the night they came back to Katupat in terror. The sides of the cave had started moving in and out when they tried to sleep. He told another story from before he was married, describing how he had once hid himself under the floorboards of a boat to make a secret rendezvous with his future wife. Another time, a girl who was five months pregnant was trying to date him, and he was accused of being the father. "Yes," he said, "I was there. But her stomach is a five month stomach, and I was only there since a month ago."

The pongkat is more than a rational economic strategy in Umar's life. It is a time of work and adventure when affective histories of travel are made, told, and re-told. Stories of pongkat travel are an important genre forming the basis for song and narrative at home in the village, even for those who have stayed behind. Umar explained that the more boats there are on a pongkat, the more enjoyable (*ramai*) it is, and the more stories there are to tell afterward. He said he had been out with as many as thirty other boats at one time. They would anchor offshore from the villages and girls would row out with coconut cookies to sell to them. Narto told me he always trades a bit of his catch to buy their cookies. I myself once accompanied a pongkat with seventeen boats on the north side of Togean Island where I was sure every fisher had gambled away his profits playing cards before returning home. The pongkat is a time of romantic adventure, flirtation, and possible riches if only the right sea cucumbers are found. Even I had acquired my own stories of swimming in the night sea surrounded by handsome men looking for precious treasures on the rocky bottom.

When it was nearly dark, the boats picked up anchor and we began to move slowly along the shoreline. As we traveled in search of a new reef to fish, the pongkat day began anew—but this time, in a new location. People who regularly collected sea cucumber moved between places where they knew certain species were abundant. Umar said to me that it wouldn't be good to fish out all the places near the village but that the variety of collecting places was limited by the distance one can reasonably

row in a night. To pongkat means to have access to remote fishing grounds, both on distant off-lying reefs and on coastal shores, spreading ones' use thinly across the resource. Umar could articulate explicitly the way the pongkat conserved sea cucumbers, while other fishers sometimes simply took mobility for granted. Like anywhere else, preservationist attitudes and knowledges are not distributed at the community scale.

Exclusionary Boundaries

I once spoke with a biologist visiting Susunang village who explained to me that since Sama people are "sea nomads" and are always moving from place to place, they can't possibly care about the particular location they happen to find themselves in at a given moment. His implication was that the people of Susunang would not protect the Togean environment because they were just going to move on. The visible world is said to provide a window on universal human reason, yet this reliance on the visible tells us nothing of the magical powers of representation to form perception. My conversation with the scientist had taken place in the house of Puah Kepala, the head of Susunang village, while we were sitting on a bright red sofa which itself rested upon a floor made of polished cement. We were drinking from his wife's Maripa's precious tea set, which was otherwise kept behind the glass of the hand-hewn wooden cabinet her brother Samal had made for her. The house was surrounded by village land that had been painstakingly reclaimed and built upon by the people of Susunang over the past eighty years. First the furniture, then the cement, and finally the house made from forest timber all seemed to magically disappear from sight as the scientist told me his story of Sama impermanence. The roof made from sago palm leaves and rattan from the garden were the last to go in my imagination as I listened to him explain marine nomadism.

Pongkat travels brought Sama peoples' collecting practices into conflict with biologists who were convinced that a circumscribed and bounded resource control, not movement, was essential for maintaining biological diversity and resource abundance. In the project of spatial planning, IFABS scientists sought out knowledge of property rights and traditional exclusionary practices and, when they didn't find them, worried about "open access." Umar's collecting involved wide mobility in search of tradable sea creatures, whereas biologists' ways of saving nature involved territorial segmentation, and bounding of people and natural resource space. But Sama people by and large do not exclude based on territory, nor do they tolerate being excluded themselves.

White Point Reef and other collecting locations and curing beaches in the Togean Islands were incrementally being restricted by state schemes to bureaucratically manage public and private land and marine tenure. Responding to what was understood as a problematic openness in Togean resource access, biologists also maneuvered to divide and delimit local resource space into use and non-use zones. Umar explained to me how people from other villages were entering the trade because of the relatively high price of sea cucumbers, and had moved into waters where Sama people ordinarily pongkat. Further, the provincial government had gazetted village borders with modernist linearity and village heads had the right to tax outsiders for harvests within these lines. Village leaders from other ethnic groups had learned to exclude Sama fishers by raising their tax, making it unprofitable for anyone from outside the village to collect there. The once-fluid intervillage relationships around collecting practices became rigid as village leaders, who were little accountable for funds, set and collected harvest taxes. Umar and his colleagues tried to avoid the tax and, resenting the new exclusionary borders, were forced to fish in increasingly tight spaces. I myself had participated with Umar in nighttime raids in the waters of other villages to avoid the tax.

Exclusionary rules had also created new ethnic tensions, and served to invent competitive harvesting practices along ethnic lines. As Sama people became the enforcement targets of natural resource regulators, people of other ethnicities realized that their own practices of reef bombing, cyanide fishing, and other prohibited activities would be ignored if they blamed Sama people. For example, when Malenge Island people (who are primarily of Togean and Saluan ethnicity) thought I was a biologist, they talked about how irresponsible Sama were. But when they perceived me as a friend of Sama people, they admitted Sama were hard-working people and they often worked alongside them in marine product trades.

Commercial enterprises were affecting coastal and forest access for all Togean peoples. Near the newly developed foreign-owned and operated pearl oyster companies, or near tourist resorts, Togean peoples were unable to collect land or sea products. At night, pearl oyster farm areas were swept by searchlights and patrolled by speedboats to keep out imagined pearl thieves. As commercial sites expanded along Togean shores, sea cucumber collecting sites disappeared. Mangroves that provided cooking wood were cut, and sea cucumber curing beaches and anchorages become unavailable as well. On Pulau Angkayo, police had forcibly destroyed Sama peoples' temporary collecting shelters to make way for a prospective tourist resort owned by the son of a former governor, and rumors of property acquisition by tourism entrepreneurs were rife.

Misunderstood as lacking a connection to place, and perceived as a people linked by primordial ties to the sea, many typically presumed that Umar

and his friends were the primary agents—not victims—of resource scarcity in the marine environment. It was nearly impossible for Sama fishers to communicate the idea that boundaries were harmful and that mobility was resource conserving, given the perception that their movement was uncivilized and their ongoing status in Indonesia as backward and alien mobile citizens. However, the story of the pongkat is a provocation to the idea that Sama mobility is harmful to the environment, that sea cucumber collecting is something people do only out of poverty and economic need, or that mobility necessarily indicates lack of attachment to place.

People Who Ride Along

The desire to control Malay peoples' mobility in the name of projects of state resource and population control has colonial roots, and representations of Sama people as nomadic fit into both historical and present-day ways of thinking about "mobile" peoples across Southeast Asia. For Dutch, British, and French colonial administrations, the stability and legibility of settled communities was a political, economic, and civilizational ambition. For example, Tania Li has described efforts of the Dutch to control the movements of upland peoples who, throughout the Indies, practiced swidden agriculture. Beginning in the nineteenth century, the household, the village, the land survey, the census, the map, and the school were among the technologies of rule the Dutch administration used to control peoples' movements (Li 1999; Aragon 2000:55–65). Control of coastal waters and Southeast Asian seas was also important. Eric Tagliacozzo (2005) has described the significance of lighthouses, beacons, and buoys for British and Dutch abilities to control trade, smuggling, and maritime travel. And for nearby Malaysian Borneo, Clifford Sather (1997:44–68) has written of how in 1901 the British North Borneo Company introduced boat licensing and registration in order to control marine traffic, collect tax revenues, and persuade coastal people to acknowledge company sovereignty.

Togean Sama people first came from southeast Sulawesi in the company of Bugis traders in search of turtle shell, sea cucumber, and dried fish sometime in the eighteenth or early nineteenth century. The earliest description we have of the Islands comes from the colonial record of Jan van der Wal, Captain of the vessel *Brandgans*, on a voyage to the Gulf of Tomini in the 1680s. Van der Wal describes encounters with a population of Bobonko people in Kilat bay and on the south side of Great Togean, and his account gives no indication of Sama or Bugis peoples inhabiting the islands at that time (van der Wal 1680).[2] Almost two centuries later, Captain Edward B. Hussey of the American whaling vessel *Peruvian*

wrote in his log on October 26, 1854, from a position between Batu Daka and Una Una, that the Togean Islands appeared to be uninhabited (Hussey 1855). Searching for humpback whales, to the northwest of the archipelago, he would not have observed the Bobonko settlement contacted by van der Wal in the 1680s or the village of Benteng (meaning Fortress) that was in existence by that time on the south side of Great Togean.

By 1865 however, C.B.H.von Rosenburg, Regent of Gorantalo, made it to the south side of the archipelago, and his account provides the first description of a Togean Island Bugis/Sama community (von Rosenburg 1865). He writes of Bugis people who lived at the foot of Benteng mountain and who had pushed the Bobonko people, whom he calls the "original inhabitants," into the interior. The interesting thing about his observation of the Benteng community is that he fails to recognize the presence of Sama people there. He describes the community this way:

> The village of Togean lies in the small, enclosed bay, Laboean Mogon, the only anchorage for ships in the whole group of islands. The houses, built in Bugis style, almost disappear in the shade of fruit trees, and are spread out randomly, on land that gently slopes down to the bay. At low tide, a broad stretch of muddy ground separates it from the sea. . . . The Bugis settled here are in part permanent, and in part temporary (*orang manoempang*). The permanent ones have their origin in Todjo, a Bugis landscape north of Bunka on the south shore of Tomini Bay. They are ruled by a Captain and they are traders and sea cucumber fishers. (von Rosenburg 1865:120)

In the nineteenth century, as at present, Bugis and Sama people often lived in intertwined communities, and what Rosenburg surely witnessed at Benteng was a community of Bugis and Sama trading and fishing together. This is evident from his improper translation of the Melayu phrase, "*orang manoempang*." Literally meaning "people who ride along," the words strongly suggests there is a Sama community conjoined with, or "riding on," the Bugis community, collecting sea cucumber for the Bugis Captain, and living off the beneficence of their Bugis patrons. The term acknowledges the patron-client relationship existing between Bugis and Sama peoples and its foremost implication is unrelated to mobility or even impermanence. The people of Susunang today call this relationship one of older and younger sibling (*kakak-adik*). There were four Bugis trading families in Susunang in the 1990s who bought sea cucumber, shark's fin, pearl shell, and other marine and terrestrial commodities and who extended credit to any who would procure these products for them.

This does not mean that both Bugis and Sama would not have traveled in search of trade and fortune in the nineteenth century as at present; as Rosenburg himself describes it, the Bugis of Benteng were from Todjo on the south side of the Gulf of Tomini, and the Dutch administered the

archipelago indirectly through Todjo until 1906. Likewise, the Togean Sama community claims kinship connections to the peoples of Salabanka in southeast Sulawesi, they share a dialect with the peoples of the Banggai Islands to the east, and they used to pay tax to the Raja of Bone in south Sulawesi. Both Bugis and Sama communities are widely interconnected along the eastern coastline of Sulawesi. In Rosenburg's description, however, the lens of nineteenth-century governmentality led him to misrecognize Benteng as divided between permanent and temporary settlers, rather than between Bugis patrons and their Sama clients.

Between mid-century and the 1880s the Togean Islands, like other parts of eastern Sulawesi, were plagued by raiders who inhibited trade and kept Togean people clustered around the defensive fort at Benteng. While the Togean Islands appear to have been a haven for the community, the Dutch administrator Revius described the situation up and down the east coast of Sulawesi in the year 1852 this way:

> According to tradition this area used to have a very busy and lively trade in all products such as rice, iron, tobacco, turtle shell, sea cucumber, woven fabrics, etc. Many boats cruised these seas, but different circumstances such as several wars of kingdoms against Ternate and, as a consequence, piracy, the tearing apart of the tribes, and destruction, scared the inhabitants and the orang Badjo [Sama] from preparing, looking for, or planting produce for trade. All trade gradually disappeared. . . . According to rumors by people who know the Bay of Kendari, there used to be a very lively trade in this general resting place for all seafaring orang Badjo and the homeward bound Makassarese, Boeginese, and Mandorese ships. This liveliness fell into decay because of the unsafe waters and the horrible lust for power of the Boeginese who settled in that Bay and rule there. (Revius 1852)

By the 1880s, the Dutch Navy, using steam ships, had pacified the Gulf of Tomini. Van Hoevell described Benteng in the 1890s: "Lampa [Benteng] used to be a place with big Buginese houses with a fortified stone wall with lillas [cannons] to protect it against the pirates mainly from Tobelo and Galela who attacked and plundered Togian on several occasions. Because of firm action taken by the Dutch navy between 1870 and 1880, these pirates have disappeared from the Gulf of Tomini and have not been spotted since" (van Hoevell 1893).

At some point after the pacification of Tomini, Sama people began to move out from Benteng to other parts of the Togean Islands. The first Sama village was located just offshore from Benteng on the island of Pulo Anau at the mouth of Mogo harbor.[3] From there, Sama people settled at the even more ideal location of Susunang, located as it was between readily accessible gardens and water sources on the land, and the most extensive system of coral reefs in the archipelago. By at least 1929, when

the Dutch naval ship *Eriadnus* produced a hydrographic survey of the Togean Islands, the village at Susunang was substantial enough to appear on the chart, although Susunang people describe living there well before then. During the next half century the people of Susunang developed new village sites at Toani to the south of Malenge Island in the 1950s, and at Kilat in the 1970s and 1980s when several families from Susunang moved there to plant coconut.

In the 1990s there were six Sama communities in the central part of the Togean Islands, all emerging from the original village communities at Benteng and Pulo Anau. Bugis people lived in all these places and continued to facilitate flows of trade and credit. Togean Sama people also have kinship connections to other Sama communities on the north shore of the Gulf of Tomini, and as far south as Salabanka. They do not have substantial ties to the Sama communities on the north coast of Sulawesi, nor do they share a dialect with them. These histories tell us that the Sama people of the Togean Islands have long maintained cosmopolitan connections to people from other places, and their history in the archipelago locates them within wider ethnic, colonial, and, more recently as we will see, national narratives of belonging.

Chicken Feathers

While one day watching television reports concerning the ethnic fighting then occurring between Dayaks and Madurese in Kalimantan, a friend from Manado said to me, "On this issue, I'm for the Dayaks because they're the original inhabitants (*penduduk asli*), and the Madurese are just transmigrants." Looking to the idea of "origins," many elite Indonesians linked their sympathies and their understanding of rights to what they conceived of as the primordial influence of place on society: an essential quality of people is to be local or not local, and mobility decreases cultural and political legitimacy. Although Sama have been described as Southeast Asia's sea nomads in the colonial record, in scholarly literature, and in popular discourse across the Malay world, floating mobility and alien citizenship also comprised an ethnic configuration with norms and forms specific to the Suharto-era developmentalist state. Elite discomfort with mobility is widespread, and many middle-class Indonesians, taught that mobility is politically destabilizing and un-modern, expressed considerable ambivalence toward the idea of human movement.

Just as settlement was preferred to mobility in this developmentalist aesthetic, wet-rice agriculture was believed superior to other forms of rural livelihoods including swidden agriculture, fishing, collecting, and hunting. Christopher Duncan writes, "In Indonesia, the government

ranks indigenous groups on a scale based on agricultural techniques and settlement patterns: the lowest being nomads, then shifting cultivators, and then settled swiddeners (those who move their fields but not their homes), all of whom rank below wet-rice agriculturalists" (Duncan 2004:5). Anna Tsing, likewise, describes how in the 1980s the Department of Social and Political Affairs classified the lowest type of "isolated population" as nomads: "Populations whose livelihoods and living places continually shift and whose living conditions and means of subsistence are still extremely simple. They still live by hunting, fishing, and gathering forest products; their communications with the outside world are almost non-existent" (Tsing 1993:155).

Within these bureaucratic articulations identities such as "nomad," "farmer" or "fisher" were proposed as fixed and distinct social spaces. In Susunang, Puah Kepala told me that the village had been invited to participate in an American foreign aid project called "Agricultural-Based Area Development." This was "foreign aid worked out between the presidents and ministers of the two countries," he said. The project would give $20,000 to each Togean village, but there would have to be a "regulation" (*peraturan*) to it: all people would be grouped according to whether they were "fishers" or "farmers." No one was allowed to be both a fisher and a farmer, and, after the process of classification was established, no one would be allowed to switch categories. Susunang people understood very well that "farmer" was a more valued social identity than "fisher," and they often claimed to me proudly that their children would only farm when they grew up.

For those without intimate experience with the sea, land and water can seem binary entities. The land is start and finish; the sea is a way to get there. The land represents the rich world of human history and domestication; the sea is a temporary pathway. But in many places with watery histories—the Mediterranean, the Pacific Islands, the Netherlands—water and land join together in configuring senses of place. In the Togean Islands, sea and land were interconnected aspects of Sama worlds. In the case of Sama people, their identification with marine nomadism meant that they were often perceived as lacking in affiliations with Togean landscapes. But this assumption ignored the anthropogenic nature of Togean upland forests and lowland swamps.

The islands surrounding Susunang are fringed by mangrove trees that straddle the line between sea and shore. Like a metaphor for the Susunang community itself, mangrove trees are soaked in salt water yet also put down roots deep into the soils of the strand. Behind these groves, Sama people had turned the land into a place productive of memory and economy. Catchments with connected bamboo pipes funneled fresh water to the shore. Channels that were cut through and behind the mangroves

provided routes to the land, and passages for boats to avoid wind and waves. Vegetable gardens, fruit trees, and yam gardens intertwined with commercial coconut and clove stands, sago swamps, and cultivated stands of nipa and pandanus. In the forest, rattan and bamboo were encouraged to grow and trees were harvested for canoe and house building. Trails reached tendril like into all parts of the forest.

The most prolific use of Togean landscapes is for production of a staple food starch called sago. Also the name of the tree itself, sago involves a particular way of domesticating the landscape. Stands of sago palm, called *gonggang* (S), grow in naturally occurring muddy areas, or are made by damming rivulets to inundate a patch of low lying land. While from a boat on the coast it was difficult for me to pick out gonggang, nearly every available low-lying place around Susunang had been converted into these swamps. There were forty-two named and individually owned gonggang along the coastlines within several miles of Susunang and Pulo Anau, and many Sama men and women who worked with sago could repeat their names and locations in an order that traced the bends and folds of the shores of Togean and Talatakoh Islands.

One time I convinced Mbo Poteiang and her husband Mbo Panai to take me with them to make sago. At the shore we climbed out of our boat, picked up our tools, and made our way up the hill to the gonggang they had named "Chicken Feathers" (*Bulu Manoh* [S]). Sago production involves travel to and through the gonggang swamp, and harvesting the palm requires many intricate movements that only locally experienced bodies could manage. Being foreign to this way of knowing the land, I could only observe and hope that my shoes would not be sucked off my feet by the mud.

I watched as Mbo Panai, to begin the process, picked up his axe and felled a ripe sago tree. Although the difference between a sago palm and a mangrove tree is visibly obvious, Mbo Panai also knew this difference in his arms and legs: the mangrove jolts the body's frame when hit with an axe blade, while the sago palm's soft trunk sucks the axe in. Mbo Poteiang and Mbo Panai worked the trunk for three full days, arriving at sunrise and returning in the afternoon each day. Panai's hands and arms flew as he flaked the pith off the inside of the sago trunk. The sago fibers slowly turned from white to red as they hit the air.

While her husband chopped at the tree, Mbo Poteiang built a sluice to strain the starch from the sago pulp. With her fingers and an awl, she sewed an old rice sack to the rounded end of a branch. Then she used the shell of an old tree trunk to make a catch basin to contain the wet sago mixture. Once Mbo Panai's pile of white inner fibers built up into a mound, he carried them to the sluice. He then planted his feet, bowed his back, lifted a bucket of water—limbs suspending the container on their

Mbo Panai and Mbo Poteiang in the gonggang, by Celia Lowe.

A-frame—and poured the water over the sago pith. He kneaded and squeezed the pith to force its starch through the rice sack. The starch flowed into the basin and, by the end of the process of separating starch from pith, his clean fingertips and wet muddy toes had crinkled.

Sago production contradicts the common idea that Sama are afraid of the land and continually leave one place to colonize another. Sama mobility in relation to sago production was something quite different; it was a circumscribed movement in and around the sago trees that created familiarity and established ownership in the landscape. Carl Sauer, the Berkeley geographer, might have called Sama people's sago-making "the appropriation of habitat by habit" (Sauer 1981:359). My experience with Mbo Poteiang and Mbo Panai in the sago swamp allowed me to think about mobility in a new way. Rather than focus on the wide-ranging mobility of a putative nomadism, one might focus on the micromotilities that form the intersection of places and bodies. While to fell a tree or maintain a sago gonggang was a manifestation of Mbo Panai's being on this place he called Chicken Feathers, to know the difference between how an axe hits a mangrove or a sago palm was also an inscription of place on Mbo Panai's body.

From the perspective of species inventory, the gonggang held a greater diversity of life than just the sago palm. Mbo Poteiang noticed a *gogomi* (S) beetle feeding on the sago pith. She told me how this beetle is only found in freshly ground sago and proceeded to name for me the other

kinds of bugs she could think of, including the sago worm. Then she picked the beetle up and tied a string to its leg to bring it home for her grandson to play with. Later we gathered mushrooms growing out of a pile of spent pith. We also collected the leaf tips of a vegetable she called *gangah pako* (S), a green and red plant that grows at the edge of the sago swamp. Sago was responsible for the emergence of a diverse ecosystem.

Mbo Poteiang and Mbo Panai were reluctant at first to take me to work sago with them. "We do work like this because we are poor," Poteiang told me. Government officials and other glamorous outsiders who viewed sago as a sign of savagery often traveled through the Togeans, teaching people that store-bought rice was a more civilized thing to eat. Before they came, sago was just what people ate; now it is what "poor" people eat. Likewise, the Togeans used to be a place for people; now they are a place for people classified as "poor." Sago consumption and production imbued Sama identity with a derogated positionality within hierarchically joined regions. Freshly made sago, still part of the tree just the night before, is much tastier than the poor quality government rice (*beras bulog*) brought by traders to the Togean Islands, however, and it is rosy in color and sugary in scent. Raw sago can be fried with coconut to make a chewy pancake, or stirred with water, lemon, and chili to make a dish like matzo ball soup. Eaten with grilled fish, it is a wonderful meal and, unlike rice, it doesn't require fresh water to cook or much wood to burn to make it.

In the IFABS ethnobiological project, the scientists overlooked the sago palm entirely, even though it is the most important economic plant for Togean Sama people. Despite the idea that the scientists were laying claim to the methods of anthropological ethnoscience, the Sama word for sago, *ambuloh*, and the sago palm landscape, the *gonggang*, went unnoticed as linguistic clues to the different interpretation and value Sama people gave to the plant. Rather, the "national" lens, through which rice is the only recognizable staple starchy food, overcame the "scientific" lens through which language is said to be a key to cognition and understandings of the natural world. While Mbo Poteiang could not escape the sense of her subordination to people who produced and consumed rice, her resistance to these relations was embodied in her persistent inscriptions on the Togean landscape through sago production. It seemed as though she was trying to impress me with her knowledge of ideas from other places when she would tell me, "We poor people have to eat sago." But I knew Mbo Poteiang loved the taste of sago more than rice, and she saw making it as cool, clean work—not like fishing under the searing sun. Turning the table on outsiders, she would tell me, "If you aren't from this place, you aren't *capable* of eating sago."

From Heteroglossia to Polyglossia

Words, mainly nouns, have been used by American structural anthropologists in debates over social commensurability and reason.[4] By taking a particular form—taxonomy—and declaring this form to *be* science, science has been found in other places. When order is found/made in systems of naming, people are said to "think scientifically." This liberal project contains within it two moves: in the first, the world is divided thereby producing a particular category of human; in the second, it is united again creating the universal human. Thus we have "ethnoscience," a mode of inscription that produces reason's "Other," yet an Other who ostensibly recognizes the proposed universal structuring of the natural world.

Here is how this might work in the Togean Islands. When Charles Darwin first described the growth of coral reefs in the South Pacific, he classified them according to three generic types: atolls, barrier reefs, and fringing reefs (Darwin 1874). His terms derived their organizing principle from his interest in the rise and fall of volcanic islands, and his classifications describe reefs in relation to the land they border or replace. In English, we still use Darwin's terms today. Terms for reefs in the Sama vernacular, on the other hand, describe reefs in relation to the sea bottom and the water's surface. These include: *sapa lannə*, a kind of reef that rises up from great depth and can be seen at the surface of the water; *sapa timpusu*, a reef that rises up from the deep but is invisible from the surface; *sapa palantoh*, a reef that is bare at low tide; *sapa patindeng*, a reef that is never dry and rises up from shallow depths; *sapa bungeng*, a flat reef that is always dry; and *sapa bapusu*, a coral rock rising up into the air from a coral bottom.

If a sapa bapusu can have a cluster of trees growing on it while this configuration describes "land" in English usage, does it follow that taxonomical differences are superstructural to nature's fundamental unity, or might it indicate the inherent instability of nature itself? If Sama are said to "think differently" from scientists about reefs (as lexicons are supposed to reveal) does this difference reside in people or in nature? Ecological, environmental, and cognitive anthropologists interested in cultural difference have used such linguistic evidence in debates over correspondence theories of nature.

Rather than look for "special cognitive abilities," Bruno Latour recommends a move from cognition to practice:

> [A] priori, before the study has even started, it is toward the mind and its cognitive abilities that one looks for an explanation of forms. Any study of mathematics, calculations, theories and forms in general should do quite the contrary: first look at how the observers move in space and time, how the

> mobility, stability and combinability of inscriptions are enhanced, how the networks are extended, how all the informations are tied together in a cascade of re-representations, and if, by some extraordinary chance, there is something still unaccounted for, then, and only then, look for special cognitive abilities. (Latour 1987:246–47)

To follow the forms of Sama people's nature-making would mean tracking emergent rationalities and practices of thought rather than codes found in the mind. If we want to use words as our guide, we can follow these practices through proper names that lend specificity to an intimately inhabited world. "Proper names," as Mary Steedly tells us, "like the lands of Lau Cih or Kilometer Eleven, assure us that something once went on there, at that point or place in the landscape, such that people cared enough to give it a name" (Steedly 1993:44).

Two miles off the south coast of Talatakoh Island lies a series of reefs strung out to the east, leading to Tanjung Keramat on the very southern tip of Walea Bahi. These reefs, the same ones that had attracted Sama people to the Togean Islands centuries ago, were classified "wilderness area" in the first proposal for a Togean park in 1982 (Salm, et al. 1982). I learned about these reefs from Puah Hamid, who insisted that he is one of the people who knows them best. Unlike generic vocabularies produced through ethnoscience, Hamid's place-names were related to practices of work and travel. These were not transcendental vocabularies existing prior to action in the world, rather his names reflected past experiences and mobilized future actions.[5]

Reefs acquire proper names through the act of building a reef house, fishing regularly, the occurrence of an unusual event, or observing a striking attribute. When he was younger, Hamid said, he would travel with his parents on the reefs to the south of Susunang on pongkat trips catching fish and collecting sea cucumber. He would sleep with his mother and father and one or two siblings on the sole of their boat with their fish catch piled on a cloth on top of them. Wherever they went to sell their fish, he told me, people would know they were Sama because they smelled like drying fish.

An interesting feature of coral reefs is that their limits and edges are never obvious in the way those of a field, swidden, or forest path might be; reefs are mainly visible by looking directly down on them from above. This attribute reveals how truly remarkable the detail was with which Puah Hamid remembered a familiar submerged geography. He described and named for me the reefs to the north, south, east, and west of Susunang. I present here only a piece of it, starting with *Sapa Ua Sidur* in the west and ending with *Tanjung Keramat* in the east—a stretch of five nautical miles.

Sama	English
Sapa Ua Sidur	Father Sidur's Reef
Sapa Alo	Atoll Reef
Sapa Gussoh ma Buka	Open Sea Grass Reef
Sapa Matilla'	Bright Reef
Sapa Buntar	Round Reef
Sapa Aloang	Has a Lagoon Reef
Sapa Kulapei	Has Kulapei Fish Reef
Sapa Sahə'	Sahə' Setan Reef
Sapa ma Rintas	Sharp Edged Reef
Sapa Mbo Mawar	Mbo Mawar's Reef
Sapa Basar Batu nə	With the Big Rocks Reef
Sapa Pai Najə	Papa Najə's Reef
Sapa Si Datei	Si Datei's Reef
Sapa Banderə	Flag Reef
Sapa Pai Jai	Papa Jai's Reef
Sapa ma Taha	Long Reef
Sapa Mbo Junung	Mbo Junung's Reef
Sapa Tarias	Tarias Reef
Sapa Ko'ko'	Ko'ko' Reef
Sapa Matilla'	Bright Reef
Sapa Aloang	Has a Lagoon Reef
Sapa Karanganang	Coralhead Reef
Sapa Boe Disalu	Fresh Water Reef
Sapa Bajangang	Bajangang Reef
Sapa Mbo Karanjah	Mbo Basket's Reef
Sapa Ua Jabir	Father Jabir's Reef
Tanjung Keramat	Spirit Cape

Hamid insisted that in order to know these reefs, one must have experienced the pongkat; reefs derive their names from these travels. *Sapa Pai Najə*, for instance, is named after Najə's father who built a reef house there. *Sapa Banderə* is named after a flag marker used for navigation that can be seen while fishing, and which was inscribed on the Togean Island hydrographic chart by the *Eriadnus*, the vessel J.H.F. Umbgrove traveled aboard while making his coral inventories. *Sapa Tarias* is named after the *tarias*, a kind of fish often caught there. *Ko'ko'* is the name of a kind of Satan that will pull out your eyes. Don't utter animal names here, Hamid warned me, or the *ko'ko'* will come. *Sapa Karangang* is named for the protruding coral rocks you can dry fish on. *Sapa Boe*

Disalu is near to fresh water on land. *Mbo Karanjah* was a foreigner, probably Dutch, who put up a basket-shaped navigation marker on that reef; this marker also appears as a fixed permanent marker on the *Eriadnus'* chart. And *Tanjung Keramat* is a Melayu name; a *keramat* cape is a location of dangerous spirits.

In much of the anthropological literature on place, supposedly bounded stable ethnic groups share putatively unified and coherent senses of place. For example, Steven Feld (1982) describes the way "Kaluli" sound expressions embody "Kaluli" sentiment, and Keith Basso (1996) describes "language and landscape among the Western Apache." Yet, as Gupta and Fergusson write, "Studies of ethnographic writing have revealed the apparent boundedness and coherence of 'a culture' as something made rather than found; the 'wholeness' of a holistically understood object appears more as a narrative device than as an objectively present empirical truth" (Gupta and Fergusson 1997:2). We can see this in the names Hamid offered.

By using proper names for reefs, we could not answer the question, "What do Sama call it?" Unlike names on a map imbued with permanence and universality, Sama place-names were continually in process, and different individuals used different place-names depending on where they had traveled and what they had done there. Hamid, for example, used the name "Father Sidur's Reef," while others in Susunang referred to the same reef as "Rizal's Reef" since Rizal had recently built a fishing house and had begun to fish there regularly. New travels and experiences reinvested places with new meanings, and the names of places could change quickly over time. When Hamid told me he knew the reefs best, he was invested not in the stability of his inscriptions, but in relating to me his experience.

Working to conserve Togean nature, scientists' interest in Sama people stemmed from a belief that damage to the marine environment might be linked to Sama ethnic identity. Although blast fishing, for example, had been introduced to Southeast Asia by American and Japanese soldiers during World War II, it became associated with Sama ethnicity in particular only in the biodiversity moment. Moreover, the logic used to link Sama ethnicity to environmental degradation was flexible and arbitrary. Rather than believing Sama ethnicity was the cause of environmental degradation, some biologists had the inverse theory, that Sama people destroy the environment because they have weak or insufficient ethnic affiliations. Dr. Supriatna once told me, "There are different tribes here and the tribes have different perceptions about natural resources. Some have *tanah adat* [traditional lands], others don't. Or maybe *tanah adat* is finished. It doesn't exist in Kabalutan [Susunang] because there is mixed culture

there." The idea that Sama culture was "mixed" had first appeared in the IFABS ethnobiological study: "Several groups of people from Bugis, Gorantalo, Buton, Saluan, and Banggai shared the island settlement. By now, the Kabalutan ethnic groups have become Bajau Sama, which means people who live a similar life style and together live over the water. The concept of 'sama' (same) arose after they were integrated with the other ethnic groups that worked together to develop the coral islands as land that could be inhabited" (Yuliati et al. 1994:8).

This analysis is a misreading of the fluidity of ethnic identities across Indonesia, however. Certainly many people lived in Susunang who had connections through marriage to people of other Togean and mainland Sulawesi ethnicities. For those ethnicities that were unmarked, like Kaili or Saluan, people tended to merge into Sama identification within a generation. For ethnicities that were marked in some way, such as Bugis, Arab, or Chinese, these ethnic identifications could last much longer. Movement in and out of ethnic affiliation in Indonesia is not a new phenomena. The historical flexibility of Sama identity to absorb and dispel members is indicated in a passage from Revius written in 1852: "There is only one trader that still gives the Badjorese indefinite, sometimes even years of credit. This is the very well known, even in Makassar and Ternate, Poa [Puah] Abang, who usually comes to Ternate to trade every year. This man is like a patriarch to the vagabonds and used to be a Badjorese himself. Through constant hard work and diligence he has become a rich man but he still keeps in touch with the orang Badjos" (Revius 1852).

Nevertheless, ethnic hybridity was sometimes understood as a strain of weakness running through the Sama community. This was especially evidenced in Sama language, where "heteroglossia"—the blending of language vocabularies and styles across indistinct borders—was deemed inferior to "polyglossia"—the codification of clearly bounded languages. Henk Maier has described how in Dutch colonial philology, "scholarship wanted order, constructed grammars and dictionaries—and concepts like borders and restrictions were introduced. Cultural definition and societal segregation were the result. Heteroglossia was transformed into polyglossia" (Maier 1993).[6] Similarly, the ethnobiological survey of 1994 describes "original Bajau language" (*bahasa Bajau asli*) as pure, while contemporary "Sama language" includes borrowings from Kaili, Saluan, and Bugis. Lack of linguistic purity was read here as a deficit, one that hindered the ability of scientists to collect "names for things" and produce indigenous knowledge. Yet, for Puah Hamid, the act of naming *is* an act of knowing.

The Strait

Sama peoples' nature-making projects cannot be read off of the text of Togean land or marinescapes, and objects themselves never structure the value, meaning, or form of the observations that allow us to perceive the world around us. Like biologists who hunt for new species by hiking through the forest leaving biodiverse nature in their wake, Sama natures are similarly produced through travel and mobility. Just as I described scientists and science, culture and nature, and species and genus as the outcomes of scientific labors, Sama identities and Sama natures were also the result of such productive work. I shared such a moment in the construction of nature and personhood on a trip through the strait between Talatakoh and Togean Islands with Mbo Dinda and Mbo Lilla.

Starting early in the morning one day in 1996, Mbo Dinda, Mbo Lilla, and I paddled their wide-bottomed canoe (*soppe* [S]) north from the village of Susunang toward Malenge Island. Rather than ride in the village head's noisy motorboat, a rattling shell he had appropriated from a village aid project, I had asked my fictive grandparents, Mbo Dinda and her husband Mbo Lilla, if they would be willing to take me to Camp Uemata. In the quiet of the day, I knew we would be able to tell stories. They had agreed and, immediately upon entering the calm of the strait, Mbo Dinda began to narrate our passage by pointing to places we passed and describing the details of particular locations. Her stories, evoked by our passage, were spun around the intersection of plants, animals, and the physical properties of the strait.

"When I was younger, that is where we would pull sea grapes from, from between those roots there," she said. If you can imagine a fluorescent green brachiated seaweed with tiny polyps on it, that is what a sea grape is, and Mbo liked to eat them with lime and chili. "And over there," she said, "is especially good for mud clams." These clams are the same Manila clam that has been imported into the Pacific Northwest of the United States. To get at the clams, Mbo Dinda told me, she would squeeze herself in between the sharp and thorny points of the mangrove roots at low tide, and reach into the mud. She scrunched up her face and hunched her shoulders together, trying to show me how little she could be, as if she were pressing her way between the roots.

After more paddling, we came upon the entrance to one of many channels that people had cut over the years leading into and through a mangrove forest. These passages are only accessible at high tide, and Susunang people maintained them for easy access to the land and a place to paddle that sheltered them from wind and waves. Mbo told me how in World War II the entire village had hid in the mangroves when the Japanese

soldiers arrived. They had lived there for months. All the children turned red from the tannin in the mangroves that rubbed off onto their skin as they played around the trees. Mbo Dinda had married Mbo Lilla during the war. "We didn't get to have a wedding party because no one *had* anything in those days," she said. Because there were no traders arriving from Java or south Sulawesi, people had to wear clothes made from pounded tree bark. It was rough and had made her itch, she told me.

We rowed a bit further, and she showed me the place where she used to collect sea cucumber. I hadn't realized that sea cucumber were also found among the mangroves—I thought they only lived among sea grasses and along coral reefs. But Mbo told me, "No, these kind are called *boto pendaga*." *Boto pendaga* I wrote in my notebook, and repeated it aloud over and over so that I wouldn't forget its name. Later I learned that I had been paddling along chanting "trader's penis, trader's penis, trader's penis!" Because sea cucumber are a trade good, most people preferred not to eat them and to sell them instead.

When we passed a patch on the shore called Red Earth (*Tane Mire* [S]), a place where P. T. Arrow M. Gobel, an Indonesian logging company, had opened up the hillside and rolled back the forest like the lid on a sardine can, I asked them where the road went. Mbo Lilla rested his paddle to talk so it didn't slap the water or clunk hollow-sounding on the side of our boat. In the perfect silence he answered, "It's a place where they've cut down the trees." We could see the remnants of gargantuan trunks the loggers had left behind. They had come with tractors with big wheels, and with jaws that tore up the trees. Then they carried them off to waiting barges, and the trees were exported to Japan. Now they were gone—both the loggers and the trees. Mbo Lilla told me *palapi*, one of the forest's huge dipterocarps, is the best wood for building a *soppe* like his. There are only two soppe left in Susunang. Now that the large trees for boat making are gone, people use *leppa*, a much narrower canoe that has one or two boards added to the edges to increase the freeboard.

"Did the company hire any workers from the village to help with the logging," I asked? "No, no, we didn't work there," Mbo Lilla replied. Mbo Dinda told me the loggers had a house in the woods, but no one stayed in it now because ghosts had entered the forest in the loggers' wake. People were killed in the logging process, she revealed, and the ghosts were a residue of those deaths.

We left the strait between Togean and Talatakoh Islands and aimed for the eastern end of Malenge Island. We would have to cross a wide stretch of water that was open to wind and waves coming from the east. Mbo Dinda and Mbo Lilla both were nervous that in the waves we might swamp the boat far from land. We talked about the whales that entered the Straits of Malenge every April. I had seen the pilot whales diving and

breaching in the open water on a previous occasion. Mbo Dinda thought whales were terrifying creatures, but I remembered this place for the beauty of the whales and my excitement at seeing them.

Arriving safely at the far side, we passed some Sama fishers from the village of Pulo Papan. They were fishing with a net, and Mbo Dinda instructed me to ask them for fish. Since Mbo Dinda and Mbo Lilla would not be able to fish that day, they would be hungry later. They knew that the fishers would be startled by the sight of a "tourist" who spoke Sama and so would be generous. I called out for some fish, "*Melaku dayah.*" The fishers swam over and dropped several large reef fish into the bottom of our canoe.

As we traveled through the strait, Mbo Dinda and Mbo Lilla recognized places for the species that were there. Places without interesting creatures became voids in the landscape and would evoke no stories. Seaweed, mud clams, mangroves, a sea cucumber called "trader's penis," a canoe-building tree, ghosts, whales, and a gift of fish were mnemonics for things doubly passed—for what we had physically gone by, and for what was history. The inverse was also true—in passing by places, the places themselves elicited and demanded story and memory in-filled by named plants and animals. Movement was constantly necessary for making the strait a familiar place—both for anything to happen and for eliciting historical narratives of what had gone on there before. Former travels had made meaningful events, and present travels brought those events to light again. With stories of whales and a gift of reef fish, a sense of Togean place had emerged for me as well.

Without abandoning the phenomenology of our passage through the strait, we might also ask what had rendered this experience with its situated interpretations possible.[7] Notice, for example, the contrast between the pleasure that I took in seeing whales and the fear Mbo Lilla and Dinda felt at the possibility of encountering them in open water. The truth of the object was not in whales themselves but in how our encounter with the creatures was modulated by propositions of danger in nature, and the relatively better position I have been in throughout my life to recover from injury. Notice also Mbo Lilla's and Mbo Dinda's ordering of nature on the basis of trade in sea cucumbers unbounded by regional or national boundaries. Notice how the forest became filled with ghosts once the important trees were removed and sent to Japan in the global timber trade. And notice the way Mbo Dinda's recollection of her wedding day was framed through a disruption in the trade of cloth from Java, and how her wedding party was interrupted by a world war. Attention to phenomena is always someone's attention; it must always be situated historically and politically.

Tribunals of Reason

Bruno Latour has written of an asymmetry between how we view so-called "scientific" and so-called "irrational" beliefs. Where science is involved, he argues, objects and phenomena (natures) themselves often seem to be all that is needed for explanation. In other instances, however, the concepts of "society" or "community" are used to explain why people believe nonscientific or "irrational" things. Latour has turned this assumption on its head to reveal both "things" and "societies" as outcomes rather than causes. "One way to avoid asymmetry is to consider that 'an irrational belief' or 'irrational behavior' is always the result of an *accusation*," (Latour 1987:185) he argues. Unreason is the product of accusation, where the accusers stand on the side of truth, universality, and transparent knowledge which needs only objects to explain itself, while the accused stand on the side of belief, myth, and illogic which can only be made intelligible through "society."

Togean Sama people were often subjected to accusations of unreason: in the way they ate sago, in propositions of their ethnic dilution, in questions of whether their mobility would harm nature. It is not only Sama people, however, who are subject to tribunals of reason, I would argue. The moment when the IFABS scientists attempted to open up the spatial planning project to Sama participants as thinking, reasoning citizens was a moment when the hemispheric divide was subverted. But Southern scientists themselves inhabit a space where their thinking, reasoning citizenship in the world is also at stake. Under the weight of accusation themselves, it is not surprising that Indonesian scientists might sometimes become caught up in a mode of understanding Sama people as extraterrestrial others. To hold this in our minds without judgment is to understand the concept of subjectivity.

Tribunals of reason are further illuminated by theories of postcolonial identity. It is not only "society" (about which Latour actually tells us very little) that results from the accusatory process, but also identity, conceived of in a fixed relationship to reason and history. As Dipesh Chakrabarty writes, "Reason becomes elitist whenever we allow unreason and superstition to stand in for backwardness, that is to say, when reason colludes with the logic of historicist thought. For then we see our 'superstitious' contemporaries as examples of an 'earlier type,' as human embodiments of the principle of anachronism" (Chakrabarty 2000:238). In Latour's desire to closely track the empirical practices and networks that bring "societies" and "things" into being, he does not sufficiently recognize the positive effects of identity or theorize rhetoric itself as a practice. Yet talk, in the form of "accusation," has the capacity to enact "irrational people."

Rather than a form of unreason, mobility was crucial to nature-making projects for Sama citizen and scientist alike. Mobility through the upland forest was necessary for making "biodiverse" nature, just as gliding through the night sea was crucial for knowing the world of *holothurians*, or swinging the axe was required to produce the gonggang and harvest the palm. The stories of Puah Umar and Puah Narto, Mbo Poteiang and Mbo Panai, or Mbos Lilla and Dinda, give us alternatives for thinking through the relationship between Sama people and Togean nature. Sama natures are practiced, storied places where what you see is not always what you get, and where objects themselves are not a sufficient basis upon which to know or to think.

Chapter Four

ON THE (BIO)LOGICS OF SPECIES AND BODIES

Inner Heat No. 28

Indication

Overcomes internal feverish sense, halitosis, dry throat, hoarse voice, fever due to cold, incessant sneezing, swollen gums, sprue and affection of the nasal passage.

Direction

Infuse one sachet in 100 ml. luke hot water, add a bit of salt. Use twice daily, morning and evening one sachet until fully recovered.

Prohibition

While still using this *jamu* avoid chili pepper, fried peanut, blinjo kernels, or fried cashew kernels.

Composition

Nasurtii Herba 25%
Glycyrrhizae Radix 20%
Woodfordiae Fructus Et Flos 10%
Fungus Calvaticus 5% and
other ingredients to make up to 100%
—Words on a Medicine Packet

As I WAS standing in the doorway to Puah Marsipe's house one day, her son Udin fell to the ground in front of me. Udin's body began to jerk, his arms flailed, and his head rolled from side to side. His trance immediately brought on a dense crowd of concerned relatives and onlookers, and Puah Hamid, who was a healer (*dukun*), was called from next door. Hamid began to suck the spirit out from Udin's head and stomach. Udin's mother was terrified, as scared as the day her two year old got dysentery. I examined Udin's body. He was only seventeen and he looked so strong. I could not translate his movements into any familiar etiology, like seizure, and there had not been any suggestion that he suffered from "chicken disease"

(*sakit ayan*), or epilepsy. I put my arm around Marsipe and hugged her trying to assure her, though what did I know of spirit possession?

I asked Marsipe what had caused this, and she explained that Udin had killed a sea turtle. The turtle's spirit had come up onto the land and taken hold of him. People in Susunang often hunted turtles and many other sea creatures; what was wrong with killing a turtle, I asked? I expected Marsipe might have told me killing a turtle is *haram* indicating an Islamic prohibition. Sea turtles are haram because they inhabit two worlds at once, land and sea, and Sama fishers usually preferred to sell turtles they caught to the Bali Hindu transmigrants on the mainland. But she had a different explanation: "the trouble," she told me, "is that sea turtles are 'prohibited objects' [*barang dilarang*]." Her description "prohibited object" indicated a recent conservation prohibition. Sea turtles were one of the forms of marine life that biologists wanted Sama people to preserve, and Marsipe felt that their list of prohibited objects should be taken seriously.

Not everyone felt this way, however. Marsipe's husband Iqbal, who had spent much time collecting at sea, found their prohibitions amusing. He had once found a turtle with a metal band attached to its leg and the band had some writing on it that looked like an address. Iqbal laughed rather defiantly, telling me he was going to inscribe "Iqbal, Susunang Village" on every kind of sea creature. "Then I will own everything in the sea!," he claimed. Some people in Susunang also framed their objections to conservation prohibitions in terms of local rights within a subsistence economy. As Mus said rhetorically when I asked him about a conservation prohibition on taking giant clam, "if the option is going hungry? [*daripada lapar?*]."

In their efforts to limit marine resource harvests in the Togean Islands conservation biologists attempted to inculcate a list of prohibitions on the world of Susunang fishers who were stereotyped by their "wild seeking" (*mencari liar*) of sea products. As with the social production of the Togean macaque, conservation prohibitions emerge at the intersection of the idea of biodiversity and concrete histories and practices of inventory and collecting. Prohibitions, which were the intended outcome of these calculations, did not necessarily produce desired results, however. While biologists tried to fit Togean people into the conservation order by instilling moral self-censorship—"buy-in," and "belief in" the conservation mission—repressed enchantment seemed to always return. Within the Togean conservation project, enchantment and disenchantment existed in supplementary relationship; new forms of "unreason" revealed the limitations of, and aporia in, practices of conservation calculation and management.

Managerial interventions in the Togean Islands often took on unexpected form, developed unanticipated lives of their own, and missed their

mark. Aihwa Ong (1987), in her ethnography of Malay women factory workers, sees attacks of spirit possession among Malay factory women as a form of womens' unconscious resistance to the rationalizing norms of factory life and emergent bourgeois values of middle-class Malay society. Likewise, Iqbal's desire to "own everything in the sea" can be read as a refusal to succumb to the disciplinarity intrinsic in the biodiversity mission. I prefer to imagine Udin's attack, and Marsipe's interpretation of it, however, as a form of supplementarity where the enchanted returns to haunt the efforts of scientific reason. The connection between enchantment and disenchantment is "turtles all the way down," to borrow Clifford Geertz's famous phrase. Because enchanted worlds exist in inevitable relation to projects of reason, the romantic, the incalculable, the un-visible, the emotional, and the meaningful are manifest in dialogic relation to projects of order.

In the previous chapter I explored the phenomenological logics of land and marinescapes through the actions of Mbo Dinda, Mbo Poteiang, Puah Umar, and others. In this chapter I extend my analysis to questions of bodies and health at their intersection with Togean natures. Cultural studies, produced through rubrics of medicine and nature, have much in common and the range of assumptions that underlie biodiversity conservation would be familiar to any student of medical anthropology. Udin's possession by the spirit of an endangered species alerts us to the connections between nature and bodies in the Togean Islands. To both species and bodies a naturalness is posited that dissolves upon closer examination. Specific understandings of this naturalness develop within cultures of nature—generalized networks of power, discourse, practice, institutions, and objects produced under the sign of the natural—that are equally applicable to biomedicine and biodiversity.

Both biomedicine and conservation are similarly viewed as remediation for "improper" or "ineffectual" knowledge or belief, for pseudoscientific practices, and for social underdevelopment. Everyday experiences of reason in the Togean Islands were determined more frequently through practices and rhetorics of health and medical knowledge, in fact, than through species or biodiversity. Shamanism, nonbelief in germs, the distance of the Susunang community from fresh water sources, and overall health statistics circumscribed Sama identity as unreasoned and in need of formal assistance to overcome deficiencies that nurse-midwives, health inspectors, and conservation biologists all perceived as interior to Sama people as a particular kind of human.

While "traditional" healing reveals how Sama people are positioned in relation to discourses and practices of scientific reason in Indonesia, medical practice in the Togean Islands indicates the situatedness of Indonesian biomedicine itself. Just as conservation biology can be delimited through

its modes of specificity and generality, biomedicine in Indonesia—practitioners, patients, clinics, medicines, diagnoses—can be specified and linked to wider networks of knowledge and medical practice. Allopathy, as an elite form of knowing that travels, is not "universal"; rather its universality is claimed and socially articulated at particular moments in time. Sama people made that universality particular when they identified me, myself, as sign of biomedicine. Cosmopolitan Indonesians produced another universalism—the universalism of the hemispheric divide—when they identified Sama people as holding backward, ineffective, and false medical beliefs.

The (bio)logics of nature and medicine pose similar opportunities for constructing an anthropology of reason. The "world out there" was known in Susunang through its effects on and relationship with both bodies and species, and land and marinescapes regularly intersected with health and well-being. Biomedicine, with its implicit links to modernity, connected medical practitioners, conservation biologists, and others with Sama people in debates over meanings of development and national belonging. Indonesian biomedical practice itself was situated, rather than the unmarked marker of a putatively universal natural body. And questions of bodies and questions of nature, arbitrarily divided into the analytical categories "medicine" and "environment," were not thus divided in Sama lives.

Lidja

In a book with nature as its subject, the image on page 16 of two dugout canoes, scudding along amongst green islands, wind filling colorful sails, suggests a tropical *idyllium* of the type that attracts tourists to the "exotic" Togean Islands. The sense of idyll in the picture comes from a particular cultural gaze, however, rather than from the nature of the image itself. I took this picture on a day I sailed with several boats out to Pulo Kubur (Graveyard Island) and our journey was actually a glimpse of misfortune and suffering. For Hamid with his hand on the tiller, or Hatija in the bow, the scene was not a moment of harmony within an aestheticized natural world. In the bottom of the windward boat lay the body of Lidja, daughter of Hamid and Hatija, wife of Musir, mother of Amrih and a new baby girl.

Lidja, body wrapped head to toe in white cloth dressed for her final resting place on Pulo Kubur, had died eight days after giving birth to her second daughter. The infant had emerged so prematurely that her skull protruded like a crown around her deeply sunken fontanel. At the time I did not hold out much hope for the daughter, but I had no idea that these

would be her mother's last days. One week previously Mbo Dinda had called me to help a sick person. Other than my bottle of aspirin, my *Where There is No Doctor* (Werner et al. 1992), and a generalized understanding of germs and antibiotics, I felt I had little to offer to people in Susunang. When I heard that the sick woman had just given birth, I had even less confidence, never having experienced childbirth myself. But I had indeed been asked for help, and so I went.

Lidja was reclined on a mattress on the floor, her body held in place on either side by "Dutch wife" pillows. She was surrounded by family who fanned her and watched over her protectively. She smiled at me but was otherwise too weak to hold a conversation. She had fallen ill soon after giving birth. On her chest was a splotch of red betel spittle where a *dukun* had cemented in place a sliver of a cock's comb. The dukun had been there earlier in the day attempting to suck the illness out of Lidja's head, stomach, and feet; then she had whispered an incantation over a small glass of water Lidja had drunk. In Susunang, ill health was usually caused by *pongko*, creatures that attack the body and consume the heart from inside a healthy person. If the heart is completely eaten, the person will die. The dukun had placed the cocks comb on Lidja's chest so that the pongko would consume the chicken part rather than Lidja's own heart.

I had been brought in as a living sign of another medical possibility, biomedicine, and perhaps I had some knowledge, pills, or powers that could supplement the work of the dukun. I looked at Lidja with no idea of what else might be wrong with her or what I might do; I did not know much about pongko—nor could I interpret her symptoms as a sickness I knew. I didn't even understand, at that moment, the range of possible biomedical illnesses that were likely to afflict someone who had just given birth. I did not understand my role, or the urgency of the moment. I sat with Lidja for a while, gave her some aspirin, and went home again in the evening. And luckily, after six days of bed rest, Lidja seemed to be getting better. She was up cooking and she bathed and nursed the new infant. Hamid told me we would play *gambus* music in a few nights to give thanks for her recovery.

But on the seventh day Lidja fell ill again. This time when Mbo Dinda called me she told me Lidja was going to die. I panicked. What could I do when I knew so little about what was ailing her. Rather than sit there in bafflement as I had the first time, I attempted to assemble objects that would bring her disease to life for me. I began by looking seriously into my *No Doctor* book. After several hours, the book helped me to piece together the possibility that Lidja had a postpartum infection, puerperal fever. The symptoms described in the book matched what she was experiencing: fever, lower back and stomach pain. According to this diagnosis, Lidja needed massive doses of penicillin. This gave me something to try.

I went to Ondine's kiosk and I bought all of the penicillin that he had for sale. I made a decision to buy penicillin rather than use the erythromycin that I had because my book provided a dosage for penicillin and I didn't know if other antibiotics would be effective or harmful. But the penicillin that Ondine sold at her kiosk came in units of 50,000 and I read that she would need 400,000 units four times a day. That meant eight pills each time and I was afraid to give her eight pills at once. If she was truly dying, as Mbo Dinda claimed, then wouldn't people think that so many pills had poisoned her? Moreover, might not the eight pills actually kill her? I decided to begin with two an hour and then work back to one every hour. That way it would be possible for her to take thirty-two pills in a day.

But this plan ran immediately into trouble. While I was invested in the utility of the clock, people in Susunang were not used to taking medicines according to clock time. The start of the Ramadan holiday was announced according to a precise hour, but little else depended on it. When Mbo Panai showed me his watch, the hour and minute hands were egregiously incorrect, and the second hand spun around the dial freely. Panai had told me the function of this second hand was to create the watch's "cheerfulness" (*hiburan*). Time in Susunang depended on the segments of the day, such as "dawn" (*subuh*) prayers, or fishing until "almost daylight" (*hampir siang*),[1] but not on the specificity of hour or minute. I went and got my clock and put it on the wall in Lidja's house, explained the dosage schedule, gave her the first two penicillin tablets, and promised to return soon. . . .

At midnight, only twelve hours later, Lidja died. I was woken up to the sound of screaming and wailing coming from her house. Until that moment, I hadn't fully absorbed the fact that death was an imminent outcome of her illness, and I must acknowledge the privilege of my naiveté; only once before had I been present for a death. How does it happen that a body can look so young and whole, and be ready to pass away at the same time? How is it that I could be speaking to Lidja one minute, and she was dead only a few hours later? "*Pitah*"(S), Lidja had said right before she died, "it is dark." No one I met in Susunang, or in Indonesia for that matter, had the luxury of my newness with death. The word *pontianak* (or sometimes *kuntilanak*) is used in Susunang and many places in Indonesia to describe the ghost of a woman who has died in childbirth. Lidja would become a pontianak.

At daylight, Lidja's family asked me to take photographs of her corpse. It felt voyeuristic to me, but they thought of it differently. How else were they going to remember her face? Her women relatives washed her body, inside and out, before wrapping her in the white burial shroud that Mbo Lilla had been saving for his own passing. They left her face uncovered

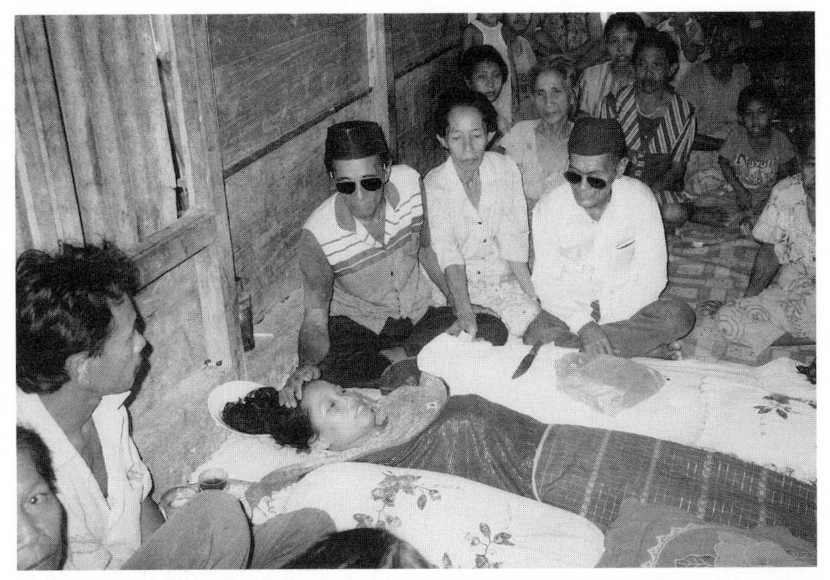

Lidja, by Celia Lowe.

for the pictures. These would be the only remembrances they would have of Lidja. It was then, after taking the pictures, that we all got in canoes and headed for Pulo Kubur. Lidja was buried before sunset according to Islamic convention and her grave was covered in beautiful frangipani flowers.

Instrumental Reason and Its Positivities

Recent cultural studies of science and technology argue that we do not want to write off a biophysical world and its effects too quickly. For example, we do not want to deny the possibility for antibiotics, immunizations, or public health programs to increase health and decrease suffering. We would not want to underestimate the power of a mimetic representation of "germs" to assist in combating Lidja's puerperal fever, or the biochemical effectiveness of the right antibiotic to have intervened on her behalf. Likewise, state-sponsored programs often can and do provide managerial and bureaucratic interventions that increase well-being for their citizens. Accounting, enumerating, organizing, surveying, monitoring, gazetting, distributing, in short "seeing like a state" (Scott 1998), all have the capacity to improve life chances for any of us.

But is instrumental reason alone what was required in Lidja's case, and is well-being the likely outcome of governmental intervention? Why do interventions made in the name of reason so frequently fail to meet their

targets? It appears there is something lacking in instrumental reason in-and-of-itself that makes it hard to argue that biological science, biomedicine, or bureaucratic rationality alone will solve the problem of early death in the Togean Islands. Moreover, in the teleology of "development," reason is expected to replace enchantment. Max Weber, Theodore Adorno, Michel Foucault and many other thinkers have argued that conditions for humanity can be substantially impaired by instrumental reason. Must reason drive out enchantment in order to operate on well-being, or are fantasies of an enchantment at peace with disenchantment only possible from the privileged position of inexperience with death?

Jane Atkinson (1996) raises the question of the capacity of instrumental reason to improve life chances in regard to her studies of Wana peoples' childbearing, life, death, and suffering in the central Sulawesi uplands not far to the south of the Togean Islands. Atkinson has explored her own dreams of immortality as somehow linked to her cultural imagination of the infinite reach of science and technology: "I imbibed a sense of progress and boundless possibility with my infant formula. I can recall brooding as a child over the death of a grandparent, then comforting myself with the thought that when I was old enough to face death as an imminent possibility, science would have conquered the problem. Fantasies of immortality are certainly common enough in the world. That I pinned my hopes to science and technology is what intrigues me about this memory" (Atkinson 1996:166).

Atkinson reflects on the unparallel life chances she has shared with her Wana friends, especially Wana women who would die all too frequently from puerperal fever and other outcomes of childbirth. Like the women of Susunang, Wana women in the 1980s were acutely aware of the dangers of giving birth. Similar to the requests I received from Sama women, the women Atkinson knew asked her for birth control. While they wanted children, they didn't particularly relish the idea of pregnancy. In a retrospective look at her ethnography of Wana shamanship, Atkinson counts all her Wana friends who have died. Of thirty-nine married adults whom she had known while living in the mountains of Wanaland in 1977, nine of the women and six of the men had died by the time she returned in 1990. Moreover, the number of men who were widowed and remarried (to mostly younger women) was thirteen, while the number of women she counted who had been widowed and then remarried was only four.

Atkinson uses these statistics to explore the question of instrumental reason, and concludes that government intervention would help to better health conditions and save Wana lives. It would be better for Wana people to be relocated to the coast where government services could reach them, she argues, although she clearly identifies the choice between Wana autonomy and Wana health as a trade-off. "The price of government-provided

healthcare would be a loss of political and cultural autonomy," she writes (182). Since Wana have refused in the past to move on their own, a certain amount of force would be necessary: "A defender of local autonomy, I am forced to concede that improving the standard of living for upland Wana would likely require coercive measures to insure that young people acquire some of the cultural capital that they can parlay into economic prosperity outside of their homeland. Such developments are underway in the region" (183).

In addition to her hopes for state intervention, Atkinson suggests that methods of scientific calculation would also be useful. Specifically, she regrets her own use of narrative analysis in her ethnography to the exclusion of a biodemographic approach. She writes, "I wanted to sound a somewhat different chord by stressing the value of precisely those research tools that many feminists, poststructuralists, and cultural anthropologists have derided as 'masculinist,' 'impersonal,' and 'objectifying' technologies of imperialist Enlightenment science" (184). While aware that a rationalized and bureaucratic intervention would contradict what many Wana people deem to be their own interests, she still finds hope in technocratic solutions to the problem of health and well-being in the Wana highlands: "Apart from castigating pharmaceutical companies for placing a higher priority on more lucrative afflictions such as cancer and coronary diseases, than on tropical diseases, it is a stretch to identify agents to blame for the physical suffering I've seen in eastern Central Sulawesi. When I do, I find my democratic principles in conflict with my desire to see the Wana conditions bettered" (181).

By coming down on the side of a Darwinian sense of survival over a Nietzschean "will to power" (both human impulses first described in the nineteenth century), Atkinson is proposing not only that this is a reasonable exchange, but that in balancing bare life chances against meaningful autonomy the likely outcome of a benign coercion would actually be improved health and well-being. But does instrumental reason necessarily direct its energies toward "life chances," or might Wana people find themselves with neither autonomy *nor* life chances?

Like Wana, Sama people have resisted regional government plans to move the village of Susunang for the sake of health and political order. Many Sama people question whether government interventions on behalf of health care or taken in the name of order, reason, science, or governmentality would improve their life chances or are even designed with their well-being in mind. It is certain that these plans were developed without an understanding of the thought village founders had put into locating Susunang close to coral reefs for fishing and sea cucumber collecting.

When I was summoned to help in Lidja's care, I was really a second choice. It wasn't that those responsible for her didn't know of biomedi-

cine, or that Lidja's family was unwilling to share the space of treatment with allopathic remedies. The village of Susunang had a nurse-midwife (*bidan*) stationed there in the government clinic. The bidan came from the Balinese transmigrant community on the mainland to the south and had been trained in the state nursing program. She had been posted in Susunang by the local government and hated living in the Togean Islands. She was standoffish around the people, finding them "primitive" and recalcitrant. I had been summoned as a possible resource, in part, because the government's own representative had refused to come.

When, after the third hour of taking the penicillin I had given to her, Lidja refused more, she said to us, "tomorrow you will bury me." A cry went up and screaming people pushed their way into the small room to touch Lidja. At this point, completely terrified, I ran to find out why the nurse wasn't there. When I found her in the clinic, she told me she would not come now, but in the late afternoon after she took her afternoon rest. Full of confidence and white privilege, I berated the nurse for her irresponsibility. I went and got my friend Jana, who was friends with the bidan, and told Jana that Lidja was going to die. We had to get the bidan to come and give Lidja an injection of penicillin, since she would swallow no more. With Jana's help we spoke again to the nurse and she begrudgingly accompanied us to Lidja's house.

By the time we arrived back at Lidja's house with the nurse, a new dukun had arrived and was massaging Lidja's side vigorously. Someone had used magic to place a needle inside her abdomen. The dukun was massaging Lidja through her ribs to try and remove the needle before it could migrate to her heart. Sharp objects placed in the abdomen through magic is an etiology found not only in rural Indonesia, it is medical knowledge Togean people shared with elite Indonesians. I was told of a similar story of sharp objects magically placed in the abdomen by a program officer who worked for the Ford Foundation in Jakarta. At this moment of urgent care, Lidja's family refused to allow her to be injected by the bidan even though I assured them I would pay the fee. I was embarrassed by my efforts to get the bidan to come when they refused her help, and the nurse immediately departed in disgust. She said to me there is no point in bringing medical attention to superstitious people.

In retrospect, the midwife, not very knowledgeable herself, or with very many techniques or medicines at her disposal, did not inspire confidence in the people of Susunang. Moreover, Hatija and Hamid did not want to leave Lidja's acute care in the hands of a hostile practitioner who might do her harm. Inherent in Atkinson's excellent questions about the relationship of instrumental reason and governmentality to well-being is the assumption many of us share that such problems exist simply due to lack of knowledge, expertise, or resources. But even in Lidja's case,

where biomedicine was available in the community, instrumental intervention alone was not sufficient to save her life. Rather than bettering life chances for Susunang women, the bidan acted merely as a symbol of governmental action. Within such a symbolic system, it is perhaps not Sama or Wana that needed to be acted upon as much as the state itself. There is little question in my mind that were Wana forced out of their mountain residence, they would end up in the same circumstance as the people of Susunang, only loosing their homelands with all that would entail in the process.

Bacteria Not Enacted

The United Nations declared 2003 the "International Year of Freshwater," and access to water is an increasingly dire environmental and social problem in many places around the world. Lack of access to clean water for drinking and sanitation is also one of the major forms that predicate how poor communities are known. Water quality as a sign of backwardness and ill health drives judgment and intervention even before any practices occur that cause "bacteria" or "dirty water" to emerge, and the issue of water in Susunang attracted a great deal of attention from outsiders. Susunang is located on a series of islets about one mile offshore from the larger island of Talatakoh and, built on tiny calcareous mounds devoid of soil, the village does not have fresh water. People spend a significant amount of time ferrying water in jugs from the surrounding land. Sama people are known for their ability to use efficiently just tiny amounts of water for drinking and washing. This does not seem unusual to me since I, myself, once used less than fifty gallons of fresh water per month when I lived aboard a boat. From my first visit to the Togean Islands, however, I perceived how outsiders were consumed by the strangeness of these water arrangements.

One day at the Susunang market I encountered a public health worker from Palu, the provincial capital. He acted as though he was happy to meet another educated person in the market, and he looked knowingly at the worm medicine I was buying. "Yes, all the people here certainly have worms," he said. "Not only that, but the water quality is clearly very poor." I had not had any problems with the water and attempted to defend Susunang's reputation. I lied and said I had never had diarrhea here. I had begun my stay in Susunang boiling my water assiduously. It was only once I realized that I was often drinking unboiled water already that I decided the water supply, which came from a spring in the forest, must be harmless. I drank the same water as everyone else for the rest of the

time I was in Susunang and only experienced a few small bouts of diarrhea, which could easily have had other causes.

The question of force in the name of health intervention came into play in Susunang in relation to water. On several occasions, the central Sulawesi government had attempted to force Susunang people to move from their offshore location to the interior of Talatakoh Island, ostensibly to be closer to fresh water. The government had even built cement houses on Talatakoh for the people. This turned out to be a financing scam designed to enrich the builder, however, and the people of Susunang never moved into them. Government officials themselves felt more comfortable on the larger islands, and it appeared to be the government's own aesthetic discomfort with the location of Susunang that was motivating the idea of relocation. In the end the government only succeeded in moving the clinic, which was located on stilts over the water when I first arrived, to some available land within the village.

The sign value of piped water was also evident in projects of economic development. The representative from the "Program for Left Behind Villages" in Susunang made a point of telling me that at his home in Poso, water came directly to his house in a pipe. From my first visit to the Togean Islands, the IFABS biologists also expressed their concern with water quality and availability in Susunang. While an experimental toilet was built for tourists in Malenge village not far from the research station, the project did not provide any water or sanitation improvements in Susunang while I was there. Rather, the scientists' discussions of water improvement in Susunang village reflected a generalized public health discourse linking water quality to social backwardness. No data were collected about the water supply. No practices were performed that enacted waterborne bacteria.[2] Yet, an object *was* enacted in the process: Sama identity.

While Susunang people's daily excursions to the water pipes onshore reflected, from an elite Indonesian point of view, an aesthetic of underdevelopment and ill health, there was no actual evidence that this way of collecting water harmed anyone. From the perspective of Susunang people, their water was clean and did not require the extra labor of harvesting mangrove wood to boil it. Living slightly offshore from the freshwater supply meant several things in terms of health and sanitation. Since there were no people or animals living near their freshwater supplies, there was little to contaminate them. Also, living offshore meant that there were not any mosquitoes, and thus no malaria in the village. Further, a brisk current swept past Susunang and so feces and dirt were carried away rapidly from the community into the surrounding sea. Finally, it was considered more advantageous to locate the village close to fishing and sea cucumber grounds than to move inland away from the reefs. In peoples' calculations

of time, labor, and aesthetics, the village was in the perfect location, and the positive consequence of locating Susunang offshore clearly augmented public health.

Left to Their Own Devices

It does not require managerial or state intervention to bring technoscience to the Togean Islands, I would argue, any more than such interventions are the path of successful biomedicine. Deep-sea diving technology is one example of a technoscientific presence that arrived in the archipelago without any such intervention. Even though some Susunang divers are able to free dive to ninety feet, they also use compressors (*komprensor*)—the same air compressors used around Indonesia to inflate car tires—to dive for sea cucumber, live fish, and pearl shell at depths otherwise not humanly sustainable. A hose attached to the compressor pumps air down to the diver, who breaths through a scuba regulator on the other end of the hose. The use of these compressors is controversial because many people have died of the "bends" (air embolism, what Sama call "*kram*") when using them. From a conservation perspective, compressor diving is also controversial in that it allows harvests of sea products to increase.

Puah Umar, who was one of the first to dive in Susunang, explained to me the devastation that compressor diving was causing in the Togean Islands. Umar had been diving for a decade since a Javanese trader first brought diving equipment and gave it to Sama fishers to collect pearl shell and sea cucumber for him. Umar had recently given up diving, however, because there were many diving accidents. One of the people who had died was Hamid and Hatija's son. The trader had given their son the equipment and was going to pay two dollars for each pearl shell he retrieved. Their son was diving near Wakai with several friends when the accident occurred. He came home from diving and was dead by evening. Hamid had already paid the bride price for his son to marry a girl from Pulau Papan. No one wanted to ask the girl's family to return the bride money because the wedding preparations were already underway and everyone felt sorry for her. If there had been another brother, Umar told me, the brother could have married the girl.

On Malenge Island, Pak Ahmad also told me of a friend of his who was diving one day when he got kram. His friend lost all feeling in his legs an hour after he surfaced and now he walks around the village on his hands. The man still dives, however, by using science/magic (*ilmu*) rather than an air compressor. He goes over the side of the boat with a rope and then uses rocks to pull him under. His ilmu allows him to breathe underwater "like a baby breathes in its mother's womb," Ahmad said.

Ahmad could not do this himself, but he could explain for me how it works. When the diver descends, he first waits for a hiccup. He must withstand (*tahan*) this hiccup and then, when another hiccup comes, he says a short prayer and the water opens in a bubble around his head. Encapsulated in the air bubble, he can stay below the surface looking for pearls for as long as he wants. "How does a baby breathe when it is in the mother's womb? That is how you breathe, using ilmu," Ahmad explained. His friend had even used ilmu to rescue an outboard motor for the Dolong police. Ahmad was afraid that his life would be shortened if he accepted the words of the prayer from his friend, and so he was unable to perform the ilmu himself.

Just like taking medicine according to a clock time, the diving that I do myself requires precise measurements. "Sixty feet for sixty minutes" is the rule of thumb recreational scuba divers use in order to avoid air embolism: more depth, then less time, less depth, then more time is safe. Umar knew about these times and tables, but he also described diving injury in the terms of a national etiology for ill health in Indonesia, the "entering chill," (*masuk dingin*, or *masuk angin*). In Indonesia, cold winds, night breezes, fans, bathwaters, drinks with ice, and other sources of cold are universally understood to create the basic conditions for illness and injury. This knowledge is shared across the class and education spectrum and is the default explanation for all otherwise inexplicable sickness, imbalance, discomfort, even death.

The deeper one dives the more likely it is that one will run into clines of frigid water. Encountering this chill, or encountering a sea satan, will equally cause kram. In chapter 3, I described how Umar understood air temperature as constant while water temperature changes from warm at night to cold during the day. With this in mind divers explained to me that one is more likely to encounter kram during the daytime since kram comes from cold water, not depth. Musir said in the day there is enough sunlight to dive to twenty or thirty armspans (*deppa* [S]), and it is very cold so one quickly gets kram there. Once, Musir came up to the surface and both of his legs had lost feeling. His legs slowly came back to life that time, but only after a dukun prayed over them and he drank some holy water. Then he got well.

Contrary to the idea of "remoteness," those technologies and knowledges that are needed for capital accumulation and resource extraction travel to "remote" locations quite easily. Anna Tsing (2004:40) has written of the "water machine" that has moved deep into the Kalimantan interior. This machine sucks water from streams and provides for the possibility of a one-man gold mining operation. It shoots water at the muddy ground with such force that it is capable of drilling a hole four-feet-deep and uncovering flakes of gold in the process. Such technologies do not

necessarily create well-being, however; they just as easily produce suffering. While traders brought compressors to the Togean Islands, they never brought the dive watches or depth gauges that would have made possible the calculations necessary for diving safely. While the water machine has produced little but debt and a ruined landscape in Kalimantan, in Susunang alone Umar could count twenty fatalities and seventeen who have been permanently injured.

It was not a dearth of knowledge, per se, that created the conditions for dive injury. Umar told me that there had been a "practicum" that went along with the compressor equipment, but that the divers were macho (*jago*) and didn't pay it much heed. Divers knew of the effects of diving on the body through a transnational theory of air embolism, through a nationally recognizable etiology of an entering chill, and through individual ilmu that could allow divers to breathe the way a baby breathes in its mothers womb. Technology does not simply drive away enchantment, it also conjures new etiological and technical imaginaries.[3] Togean divers had combined ilmu with the mechanical knowledge needed to convert a tire compressor into a diving rig and entered the deep-sea in a new way.

This ability of technology to spark invention and imagination was as true for scientists as for Sama people. In Jakarta I attended an ICDP planning meeting at IFABS, where we discussed a forthcoming workshop that would take place at Camp Uemata. The scientists wanted to spread knowledge about biodiversity conservation among Togean people, and I had been invited to contribute to the workshop agenda based on my experience living in Susunang. Since Sama people were frequently injured in diving accidents, I thought this would be an opportunity for the conservation project to show that it cared about Sama people's lives, not just about plants and animals. Diving is demonstrably safer when practiced within contexts of instrumental reason and, at the time, I believed the problem of diving in the Togean Islands was simply one of knowing about air embolism. My idea was to teach about water pressure and the bends by using a bottle of clear soda. By releasing the pressure from the bottle, bubbles would form the same way they do in the blood stream. I wanted to use this as an illustration.

Several months later at Camp Uemata I attended the workshop we had planned in Jakarta. Many scientists and community educators led discussions, and participants from all the surrounding Togean communities attended. The message of the workshop was, "we live in nature that is not really ours" (*hidup di alam yang memang bukan alam kita*), and the primary emphasis was on tourism. I was happy to see that the question of diving was on the agenda. When we came to the discussion of diving, however, the agenda had been transformed into a conversation about how to promote dive tourism and keep tourists safe. Like a prince

into a frog, the discussion of air embolism, as I had conceived of it, had been magically transformed.

I view the fact that my own attempt at rationalized intervention on behalf of Sama people's well-being was unsuccessful as exemplary of a generalized theory of reason. New enchantments arise together with efforts at disenchantment, and reason is never "pure." It always attaches to a person or group and their intentions, conscious and otherwise. In the case of the Uemata workshop, what seemed reasonable to the scientists was an intervention on behalf of tourism in the Togean Islands. Tourists do not harvest marine products. In fact, their presence is supposed to substitute for harvesting wild sea creatures. The Togean participants were taught to make coconut shell necklaces on green plastic strings to sell to tourists in place of harvesting wild sea or forest products themselves.

Such phantasmic leaps of logic inhabited the very heart of the rational scientific biodiversity enterprise, while Sama "backwardness" was said to reside in their distance from science, technology, and medicine. This distance entered into Sama peoples' own self-consciousness of their place in the world. "In America, everything is done with machines," Musir explained to me about my country. "You don't do anything by hand the way we do here," he said matter-of-factly. Despite the abundance of manual labor in Susunang, Sama compressor diving belies the existence of remote places distanced from modernity through lack of science or technology; technology follows capital, even into the most capital-poor locations. In the Togean Islands, the air pump was adopted and adapted easily by Sama fishers in order to make the undersea world and its marine harvests more accessible. What was not possible, in this instance, was for unmotivated well-being to follow upon these flows of capital, expertise, and equipment. For this, Sama people were left to their own devices.

The Python and the Cuscus

Like the use of compressors, the world of pharmaceuticals in the Togean Islands was not immediately recognizable to me, nor was it utterly alien. When Puah Iqbal decided to take the steroid dexamethosone to make himself look chubby and handsome (to me he looked simply bloated), I could recognize the name of the pharmaceutical company on the package of medicine, and I knew that this steroid had another purpose in a different context. If Iqbal's liver became damaged from the steroid as a consequence, was this a biomedical effect, or an effect of superstition? It was certainly a result of experimentation, since Togean people had tested the drug and noticed the effect it had on body weight. Iqbal's use of the

medicine he called "dexa" blurs the boundary between allopathy and home remedy.

Similarly, when I met the bidan in Jana's house treating Jana for "inner heat" (*panas dalam*), I learned that this was more than an immediately recognizable description of fever. Inner heat is a description for illnesses that exhibit a sore on the outside of the body. A herpes sore or a hemorrhoid indicate *panas dalam*, and these external wounds are a sign of internal disorder. Unlike the "entering chill," which is a universal Indonesian etiology for illness, "inner heat" is a class-bound syndrome and only some elite patients would describe their symptoms this way. Yet the bidan, who was a trained biomedical practitioner, diagnosed Jana with panas dalam and prescribed for her a dose of an appropriate pill. Through the diagnosis of the government-sponsored nurse midwife, panas dalam moved into the realm of biomedically remediated illness.

Each Wednesday morning Susunang held a market where medicines were sold. On Tuesday night, traveling traders would arrive on their wooden cargo boats. At night the boats took onboard firewood, rattan, salt fish, and other sea and land products Susunang people had to sell. Before dawn the next day they would unload rice, flour, sugar, and other food staples, plastic containers and cooking pots, sarongs and children's outfits, nails and hardware, and many other enticing items. Along with these dry goods were the many kinds of medicines (*obat*) brought by traveling obat sellers. The obat sellers set up tables in the market with boxes of tetracycline and ampicillin, vitamins of all letters, parasite medicines, antimalarials, and steroids. Many were made by international pharmaceutical companies with names I recognized: Bayer, Pfizer, Bristol-Meyers Squibb, Merck. There were other medicines produced by Indonesian companies: Kimia Pharma, Bintang Toedjoe, Dexa Medica, Konimex. And there was herbal medicine (*jamu*), some in packages imitating the pharmaceuticals, and others in bottles containing home-brewed syrups and powders in various colors and combinations.

What is it that makes any particular medicine "biomedicine," or any given practice allopathic? In the case of the pills produced by pharmaceutical companies, it could be the practices of testing and experimentation (Shapin and Schaeffer 1985) that have gone into developing their formulae, or the chemicals they contain. Across each packet of pills was written, "not to be dispensed without a doctor's prescription," so the medicines might be made "biomedical" through the regulatory practices governing how medicines are dispensed, although the obat sellers would sell them to anyone with money. Jamu, likewise, could also be considered biomedicine when it came in standardized mass-produced packets with Latinized ingredient names listed according to their percentages.

In its imbrication with pure reason, allopathy proposes a human body that responds to universal forms of medical intervention. In this configuration the idea of a "universal" biomedicine contrasts with a situated "local" medical practice, usually referred to as "traditional." But Indonesian biomedicine is also particular in the way pharmaceuticals enter into circulation, in how and what medical practitioners diagnose, in the ways patients understand bodily experience, and in what patients expect from their interactions with practitioners. Biomedicine is not a fixed object with universal properties. It is merely a form that is able to *claim* universality. Sometimes Indonesian people choose to enter into that universalism, and other times they reject it.

Each week the obat sellers used animal acts to attract people to their medicines, giving the market a circus atmosphere. One Wednesday an obat seller appeared in the market with a cat and some mice in a cage. Everyone knows that cats and mice cannot live together! How could this be, if not for the strength of his medicines? Another time, one brought a crocodile. The crocodile was as potent as his powders, the medicine man proclaimed! One day, an obat seller arrived from North Sulawesi wearing an embroidered gray suit (*kerawang*), a tie and, on his breast pocket, a badge with his photo stamped with a seal. This "license" was meant to persuade, as were the python and the small furry marsupial, a cuscus, that he carried with him in cages. He uncovered the animals in their cages and began to orate in the tradition of P. T. Barnum: "Gather round, gather round! I have already sold my medicines in Ampana, in Wakai, in Dolong [towns in the Togean Islands]. If you don't want to buy them, that's fine, others will want them. Here, just take one of these pills every day. Or wash with this soap. If you want to just look, go ahead and look, that's ok. People who know the strength of my medicines will buy them."

Narto and I were interested in what he was saying and we both were intrigued by the creatures. He took the snake and the cuscus out of their cages, and wound the python around his shoulders. Why didn't it strangle him, I wondered? The obat seller appealed to the public by invoking a connection between health and wealth, reminding us of the stranglehold the rich had on the poor:

> How many of you here are rich? You could be rich, but if you aren't healthy, what does it matter? I am not here to get rich! Only to help the people. What does it matter if you are rich, if you are sick. We know it is money that rules the world [and he rubbed his fingers together]. Just ask yourself, who is it that is in jail now? It is the people who have stolen a chicken, or just a little rice. Rich people commit murder and they still get out of jail. How? By paying! How about our former Governor!

I asked Narto about the obat seller. What was special about the medicines he was selling? Narto told me the seller was sent by the government. There was his badge, we could see it hanging on his pocket. I asked Narto again about the medicines. How were these medicines different? Narto explained it to me, "They are like all like the medicines you have, Lia. They come from America."

The obat seller continued to attract onlookers. "Gather round, gather round! This is the era of development," he cried: "Our president is the Father of Development, President Suharto, our second president. Some will be wealthy and some will be poor, but what is wealth without health! I am here for you as a bridge between the city and the village. I'm just doing a service for the benefit of others. I am here as a bridge between the government and the people."

The geography of progress that mapped onto "medicine" was the same as the one that mapped onto "nature." Sama people were positioned as keepers of false knowledge and inadequate belief in a system where facticity was determined as much by social position as by the empirical demonstrations necessary for elaborating a system of science. Contrary to the idea that marginalized people are resistant to biomedicine, Narto's comment hints at a complex spatiality to medical practice. Effective medicines come from far away, from the city not the countryside, even from America. The village will always be "left behind" in this scheme. The government is purveyor of "development," but development does not mean well-being in the obat seller's rhetoric. Development means that "some will be rich and some will be poor." Like the boa constrictor, "our former Governor" could squeeze money out of the project of development. But like the furry slow-moving cuscus, there is also a place for the people. If the people have health, they too can be happy (*senang*).

That medicines, and not the metal tubs painted with bright orange flowers, glittery polyester princess dresses, or tins of hair grease, became the material of spectacle was fitting in the context of Indonesian medical practice where biomedicine and traditional healing alike depend upon hidden powers. Once, I accompanied a biologist friend to the doctor's office in Manado, a site of urban biomedical practice, where her doctor gave her prescriptions for her illness. At the pharmacy she received one plastic bag with ten red pills, another with fifty white pills, and a third with three yellow pills. None of the pills had names, nor were the bags labeled. In urban Indonesia, biomedicine also depends on the concealed knowledges of practitioners, not on public cultures that encourage critical knowledge of medicines, their properties, and their uses. Like the dukun's secret incantation over a glass of water, my friend was expected to take her doctor and her pharmacist at face value, to be convinced by their symbolic place within universalist biomedical reason.

Healing as Care

Stacy Leigh Pigg has written, "Shamans figure in the Western imagination as objects of science, obstacles to science, and alter images of science. Shamans, taken to represent 'tradition' (or some times pre-modern worlds), have long been handy symbols in the construction of 'modernity' " (1996:161). Many people in Susunang were self-conscious about discussing aspects of health and well-being, especially spirit possession or their experience of pongko, because they were well aware of the proposed duality of superstition and reason, and on which side of that duality their community is said to lie. Biomedicine is proposed to be about objects—the body, the bacteria, the disease—which determine and motivate appropriate actions. But, as the case of water illustrates, biomedicine was also used to motivate action before any objects themselves were enacted. Bacteria were known through the same means as pongko, through language and speech acts that brought them into existence by invoking prior contexts. In the Togean Islands, the proposition of a universal medicine, founded on stable biological objects, was productive merely as a system for classifying people. It was not an effective means for improving public health.

Puah Samal told me how one has to have money to enter the hospital or else the doctors will allow you to die. After saving up what he thought were sufficient funds, he entered a hospital on the mainland of Sulawesi to repair his hernia. I heard that he died on the operating table. When I took my friend Budi to the dentist in Dolong, the dentist accidentally broke Budi's tooth off at the gum and left the fragment still rooted in his mouth. A friend in Jakarta whose family came from Maluku Province told me the reason the population was so low in Maluku was because the government family planning program (*keluarga berencana*) was sterilizing women without their consent. This strategy of biopolitical control was purportedly used on the occupied island of East Timor as well. In Susunang village a second government nurse assigned to the village had moved to Malenge Island to open a tourist hotel, where he could make more money. He would come to Susunang only if a patient's family could rent a motorboat and pay the fuel cost needed to pick him up and drop him back in Malenge.

Under these conditions Susunang people cannot be understood as a different kind of human with different kinds of "knowledge" or "beliefs" about the body. Lidja's death raises more than the question of belief in germs, or knowledge of allopathy or traditional medicine. Rather, biological and other medical practices produce subjects of health and disease through entire networks that are at once social, political, experiential,

and material. The missing links in this network—transportation systems to take people to doctors, laboratories to enact diagnoses, regulations over how drugs are dispensed—really have nothing to do with belief or knowledge. I certainly knew less about childbirth than the dukuns of Susunang, and I was unable to help Lidja despite my belief in germs. Like blue-tailed lizards, which required the laboratories of the Smithsonian Institution to turn them into new species, germs require an extensive network and a complex process of enactment to emerge as credible objects.

Rather than using "force" or "judgment," one solution to the problem of puerperal fever in Sulawesi might be to provide effective and compassionate health care and allow women to make use of it of their own volition. This path aligns with what feminist scholars have called an "ethic of care." These scholars (Gilligan 1982; Hankivsky 2004; Tronto 1993) have contrasted an ethic of care with the ethic of justice described in liberal political theory. Care depends upon acceptance of human difference and uniqueness, rather than on the figure of the "universal human," whose requirements for justice are determinable in advance. As such, care attends to the specificities of context and individual experience, and acknowledges human interdependencies in both the public and private sphere. All people need care throughout the life cycle, moreover, and so a hemispheric divide is not forced on any segment of the population. In such a world, the methods of scientific calculation and a government's organizational capacities would be put to the task of health *care*, rather than toward ensuring political order or creating the aesthetic forms of modernity.[4]

During the course of my research, I once found myself floating in the middle of the Gulf of Tomini after my ferryboat sank out from under me in the middle of the night. This, I later learned, had been caused by a lack of rationalized oversight over public transportation. The owner of the ferryboat, Agape II, had been paying the Harbor Master in Gorantalo to forego regular safety inspections. A rotten plank underneath the engine was not discovered in time, and three people died in the accident. We were finally rescued when a commercial fishing boat set its nets and pulled in three passengers. The crew of the fishing boat then gathered the rest of us out of the sea. When we reached the shore in Gorantalo, we were greeted by a phalanx of nurses who, as a form of assistance, wanted to give all the foreign passengers inoculations. I refused their offer, as generous as it was, not believing that an injection of an unknown substance with who knows what needle was any kind of remedy for floating in the ocean for five hours and losing all my luggage. And why was it that only the foreign passengers were deemed to need injections? Biomedical and technological reason may be oriented toward life-saving ends and enhance well-being, but this is not necessarily so.

It is possible to speculate, as Atkinson does, that Wana people are remote from technoscientific and managerial capacities, rather than that they lack faith in the options available to them closer to the coast. It is not as clear to me, however, that the more "remote" parts of Sulawesi really are impervious to science and technology, or that state-sponsored technoscientific management regimes really are absent. The introduction of allopathic medicines and of scientifically trained nurse-practitioners or midwives has not, thus far, been the solution to the problem of puerperal fever in the Sulawesi hinterland. Stories from Susunang are a provocation to the idea that life chances improve merely through the presence of technoscientific, biomedical, or instrumental reason. Many of us would want to refuse an injection, or choose a caring family over a hostile or unfamiliar government representative, just as Lidja did.

PART THREE

Integrating Conservation and Development

> Why should we preserve nature? For three reasons: 1. Nature is the creation of God. As Muslims, God does not allow us to harm the environment. 2. In the Government's "Broad Outline of the National Direction" [*Garis Besar Haluan Negara*], the people are given the responsibility to protect the environment. 3. For the promotion of life, the continuation of life, for the life of all earth's people, it is the peoples' work to protect the environment.
>
> —Budi, Ecotourism Practicum, May 1997

THE INTEGRATED Conservation and Development Program (ICDP) was a particular solution to the problem of nature and the human that operated at a global scale within 1990s biodiversity conservation. The idea of integrating conservation with programs of economic development was based upon two premises: first, communities that live near biodiverse nature put that nature at risk, and second, prior efforts at nature conservation failed because they did not take into sufficient consideration the economic needs of people living near protected areas. ICDPs propose that people who live in or near conservation areas have a right to make a living from their surrounding land and marinescapes, but this income should come from nonconsumptive uses of nature. "Ecotourism" was the most common programmatic solution imagined within this global framework.

While the IFABS project adopted the institutional form of the ICDP, what "conservation" or "development" would mean in the Togean Islands was interpreted through domestic understandings of national progress, rather than through an economic abstraction of income substitution or through "universal" ideas of modernity. Development (*pembangunan*) in 1990s Indonesia possessed a certain "structure of feeling," a specific discursive and interactive modality, that brought together such elements as: Indonesian class hierarchies; the role of the Indonesian state; the place of education in national development; the figure of the citizen; an assumed difference between "East" and "West"; a national conversation on religion; and other specifically domestic modes of political discourse and identification. These sensibilities made the Togean ICDP distinct from the transnational regulative discourse on the importance of mixing development with conservation.

I experienced these particularities firsthand at an "Ecotourism Practicum" conducted by IFABS at Camp Uemata in May 1997. The scientists had invited Togean Island people from four different village communities to attend the meeting. They also invited the local representative (*Camat*) of Walea Kepulauan subdistrict, and a tourism administrator from Poso to make presentations. In Susunang, however, the practicum ran into a problem at the start. I was on my way to Camp Uemata to witness the events and offered to give a ride to whomever would represent the village, but Puah Kepala had not yet chosen anyone to attend. "Why don't *you* just represent us at the meeting?," Puah Kepala had asked of me. I responded I could not represent them since I wasn't from Susunang. Puah Kepala then ordered his teenage niece to accompany me to the meeting. In selecting this young girl, he was indicating the antipathy he felt toward IFABS and their work.

I arrived at Camp Uemata with Puah Kepala's niece in time for the opening ceremony, which began with a prayer and a speech by the Camat. Then, Budi introduced the purpose of the practicum. "This work is for the development of all of us and for our nation," he explained. We would learn about tourism, about conservation, and about how to employ conservation as a form of economic development and self improvement. IFABS was not entirely in control of the agenda, I observed. It was necessary to invite the Camat and several other government representatives in order to demonstrate that this workshop was not subversive of government intentions. The practicum could not be held without the cooperation and participation of regional government representatives.

The first speaker was Pak Pujo, the Director of Tourism from Poso. He began, "There are still some people who are not responsible (*tidak bertanggung jawab*) and who destroy the environment." Tourism, he suggested, could provide a solution to this problem. We might be confused, however, or have an "unclear understanding," about what tourism actually means. "For this reason, the government has created Law 9 of 1990 Concerning Tourism." From Law 9, we should be able to understand that tourism is connected with governance and therefore is the responsibility of the state. "We live in a rule of law state (*negara hukum*) and therefore we should follow its laws. In the law, you will find everything ordered there," he explained. He told us that, in Indonesia, law begins with the President, then it descends to the ministers, next it comes to the national level, then the region, and finally it arrives at the local level of government, which includes his office: "Office of Tourism, Level Two." "Our law is based on *Pancasila* [the national ideology]," he said, "but I will move on from here so that my talk does not smell of government interference (*berbau pemerintah*)."

He continued to discuss the tourism law in order to erase our confusion about the meaning of tourism. Law 9 of 1990 defines a "tourist" (*turis*) as "one traveling, according to their free will [*dengan suka rela*], to see visitor objects [*obyek wisata*]." Therefore, he told us, a turis is one who tours. Turis come in two kinds: international turis and domestic turis. "Tourism" means everything that is connected with turis. He went on to describe the twelve responsibilities of his office and what was outside his purview. For example, the hotels with the star rating system (*hotel berbintang*) are the responsibility of the central government. But if you want to open a small hotel without a star rating, you would need a permit from his office.

While he continued to give us the definitions of the different kinds of hotels, and describe for us the appropriate division of labor so that everyone in the system would benefit, I saw the women participants sneak off to make coffee. "It is important that we all share in the tourism profits," he went on, and then he explained to us what profits should be guided to his office, and what profits the people themselves could keep. "If the system is adhered to, the tourists will bring out their dollars, the nation will profit, the local government will profit, the people will profit, and the tourists will have a good time," he explained.

One of tourism's benefits is that it can help to preserve culture (*budaya* and *adat-istiadat*), Pujo argued, and he instructed the audience to prepare for tourism by making ready an "original marriage ceremony." The Lake Poso Festival would chose one marriage ceremony from the Togean Islands next time, "if the people were ready." But it was important to find an "original ceremony," and not one influenced by Gorontalo traditions or the traditions of others. Tourists really appreciate seeing this culture, he claimed, but only if the culture develops on its own (*tumbuh sendiri*). He cited Toraja death ceremonies as an example of culture that has developed on its own.

Along with tourism's positive influences, there are negative ones. Tourism, we learned, presents many dangers. It can have "cultural side-effects" and requires particular attitudes and proper conduct. Tourism can "ruin the social contract," Pujo warned, and we must be vigilant. "Togean people practice and follow Islam, and Islam is a religion that instructs us in our daily lives," he warned. The people must follow Al Qu'ran, which means that they should not harm nature. But, more importantly, it means Togean people should not follow culture brought in from outside. "We have our own culture. Indonesian culture. *Pancasila* culture. We have to be careful. If we see something wrong, don't follow that just to get closer to our guests. We have to remember that they are from the West, and we follow Eastern culture. For example they wear those 'mini' clothes," and he drew his hands across his thighs as though he was wearing a bikini.

Again he invoked the name of God, "God does not grant us permission to harm the earth. The earth is not a gift of our ancestors, they are only letting us borrow it." The way to have successful tourism, Pujo told us, is to follow the seven important principles: (1) safety (*aman*) (2) order (*tertib*) (3) cleanliness (*bersih*) (4) freshness (*kesejukan*) (5) beauty (*keindahan*) (6) friendliness (*rama-tama*), and (7) remembrance (*kenangan*). He concluded with advice for the upcoming national elections:

> At our election (*pesta demokrasi*), it should be remembered that our development is the result of the New Order. It is the result of Golkar [abbreviation of *Sekretariat Bersama Golongan Karya*, the regime party]. When I was on the boat coming here I head the ship's bell which reminded me of our upcoming elections: One ring meant stop and three rings meant go backward [references to the two other legal political parties]. But two meant 'forward' [*maju*]! Of course it is your choice what party you chose.

I asked Budi about Pak Pujo's presentation. Had Pujo conveyed the message Budi had wanted? He explained to me that in the past the government and NGOs were "allergic" to each other. Now they try to work together. That meant the NGOs had to make concessions to the government position. They could try and influence the government just a little bit by introducing the idea that conservation is important, however. For example, he had given Pak Pujo the idea of a guide association and promised to help with planning the content of the program. He explained that the instructions that come out of the central Jakarta offices are "70 percent *Pancasila* and only 30 percent tourism." "You can't eliminate the *Pancasila* content altogether," he explained to me, "or else the local officials will just throw your work into the trash. But maybe you can shift the balance, 60/40, and throw in something about conservation."

On day two, Budi presented IFABS' own ideas for integrating conservation and development in a presentation titled "Developing the Capacity of the People." Budi began with the questions: "What does it mean to build our capacity?," and "Why do the people have to be strengthened?" The audience was slow to respond. A participant from Katupat village said, "we are fishing for the right answers first, for the answer that you want from us." Then, one participant called out, "the people need to be strengthened because they are weak." "Good," Budi encouraged, "and what are the peoples' weaknesses?" He turned to the blackboard and wrote down the responses as people called them out to him:

1. weak in education
2. weak in economy
3. weak with laziness
4. weak in communication media

Budi himself added to the list without noting it on the board: "Weak in the ability to overcome outsiders who are reef bombers and cyanide fishers."

"Why are we 'weak'?" Budi asked, and then something unexpected happened. A young man stood up and asserted, "we have a problem with government. If our government was just, we would be given access to school without school fees. Many people here are smart enough to go to school, to college even." Nervous laughter filled the room. Budi underlined "education" several times with his marking pen. "We need more education," Budi replied and then moved on to number two. "What about 'economy'?" The same participant spoke up again, "if the government organized more work opportunities, and organized our education, there would be more work and we would be able to do it."

Budi intervened, "but we aren't like a developed country (*negara maju*). The government can't give us everything. We must learn to help ourselves." Budi then asked, "what about 'laziness,' why are we lazy?" Again the outspoken young man raised his voice: "Try building a junior high and high school here. If we are sent to town far away that only adds to the costs that our parents can't afford. We loose the chance, so we become lazy." By this time, the mood was uncomfortable. Another participant jumped in: "Make the people more hardworking. If we are more hardworking, we won't be lazy." "Good!," Budi answered, and he looked around the room to see if everyone agreed. Heads nodded in agreement.

Budi next asked us to define the word "nature" (*alam*). "Forests!" "The sea!" "Stars!" "The moon!" People called out the elements of nature. Budi expanded on these suggestions. Nature is comprised of "ecosystems" (*ekosistem*) he explained, and an ecosystem is a group of "biotic communities" (*komunitas biotik*). Indonesia has forty-seven types of ecosystems, he told us, including sea, beach, mangrove, mountain, et cetera. Then he asked, "Why should we protect the animals and their habitats?" People remained silent. "Because of the food chain," Budi answered. "The plants are eaten by the herbivores [*herbivora*] and the herbivores are eaten by the carnivores [*karnivora*]." "If the chain is broken, all will be broken," he said.

He asked us to think of the trees. What will happen if we cut down trees? He was especially concerned about the Malenge forest since it was home to the Togean macaque, but Budi attempted to orient his response toward the participants' concerns: there will be erosion, floods, and the wells on the island would dry up. "The trees are an easy example," he told us, "while the sea is a much harder one. The sea is more difficult to understand because the system begins with plankton which are tiny and you can't see them." He explained to us that we could ruin the sea and damage the food chain. The question is, how to use the environment with-

out ruining it—ruining it, for example, by using bombs and poisons? "If we use a bomb or cyanide, we are quickly full one time, but later there will be nothing—there is only a momentary enjoyment. You are happy once, but the coral is broken and ruined," he argued.

He turned to Pak Ahmad and asked, "Isn't this true?" Ahmad agreed. "So do we need to bomb, or not?," Budi asked us rhetorically. "Noooo," we answered.

Part 3 addresses what it meant to attempt to integrate conservation and development in the context of Indonesian law and the "reason of state" in the 1990s. The Ecotourism Practicum at Uemata that week represented only one type of effort to integrate conservation and development. The IFABS scientists had also worked extensively on tourism initiatives over several years, including starting a guide service, building a nature trail on Malenge Island and a boardwalk through the mangroves outside of Katupat, and teaching people to make souvenirs. In the midst of these efforts a second "integration" was required: integrating conservation with the state's ideas of progress and order. In order to continue their work in the Togean Islands and to be successful in their goal of creating a national park, the scientists were compelled to do so in a form that honored the authority of the state and respected its norms and aesthetics. As a consequence the major scale at which responsibility was assumed to be located, and therefore the scale at which intervention was imposed, was the "community." While the transnational rationale of the ICDP claimed that communities are both source and solution to the problem of nature at risk, this is the same scale the Indonesian state uses to apportion blame and discipline. The overlap between transnational and Indonesian state discourses overdetermined the form of the ICDP as something that would enhance state control even as it spoke of popular prosperity.

Chapter Five

FISHING WITH CYANIDE

> Making the law work as it should in Indonesia is very difficult. The legal system is stacked in favor of the wealthy and powerful but even limited progress can have an important symbolic impact. Law enforcement has to be carefully targeted if it is not to create more conflict, however. . . . A more effective strategy, and an easier one to justify to the public, would be to target not the low-paid workers of the industry but the [business leaders] and corrupt officials who organize and facilitate it.
> —International Crisis Group, Indonesia: Natural Resources and Law Enforcement

On only my third day in Susunang, in January of 1996, I witnessed a major event in the history of Togean cyanide fishing, one of the methods of fishing that most concerned conservationists. Walking through the village that day I encountered several men arguing over some jerry cans and stopped to see what the ruckus was about. Sitting quietly at the edge of the action, I asked my friend Udin what was happening. He told me that the crew of a fishing boat had been "arrested" by the village. The boat had been fishing for live fish close to the village and was using cyanide. These were their jerry cans and they contained solutions of cyanide poison.

The captain was a tall, lean man in T-shirt and shorts who spoke with utter confidence. The conversation took place in Indonesian, the national language, so it was clear he was an outsider. The captain was from the Sangir Islands in north Sulawesi, it turned out, and his crew was from the Philippines. A crowd of men had gathered, and everyone was shouting at once. I noticed Mahmood (a Bugis trader from the other end of the village), Udin's uncle, and many other important people whom I would later come to recognize as leaders in the village cyanide fishing network. The captain bragged that he had important friends in the police in Luwuk, Poso, and other places, and the "arrest" would not stick. When the argument ended and he finally turned to return to his boat, he stopped to ask me in a swaggering voice where I was from. He took me for a tourist, and his bravado symbolized his cosmopolitan connections.

I asked Udin if we could go and see the boat. At first he said we could go, as the boat was nearby. Then Udin's uncle suddenly interrupted and told us that it was too far. Who was being protected, and from what, I wondered? Later, I heard people discussing Camp Uemata and I over-

heard someone say that I would report what had transpired directly back to Jakarta. Some people believed, early on, that I worked for IFABS even though I had tried to explain my research. IFABS scientists had been discouraging cyanide use by telling people to report on their neighbors. No one would do this, of course, but not wanting trouble for family and neighbors is different from approving of cyanide use. Udin later explained to me, "the problem is that the people here are small fishers. They have no other livelihood but fishing. They will have nothing if the reef is killed."

I heard many different perspectives on cyanide use during the two years I lived in the Togean Islands. For example, while the "arrest" appeared to be motivated by village leaders' territorial claims to fish, rather than by a true objection to cyanide fishing, many other Susunang people were sincerely disturbed by poison use. Months after the arrest of the occupants of the boat and its captain from north Sulawesi, I discussed cyanide use with Puah Jafir, a friend who had taken me fishing many times. Although Jafir had taken up live fishing himself only recently, he already noticed a change in fish stocks. "If people were using poison and my take dropped only a little, I would accept it," he said. "But I feel heartsick people have used cyanide here, and then I catch nothing at all. I have not caught a big fish in a month, so there's no point in fishing this afternoon. There won't be any results."

When I had fished with Jafir, we paddled his canoe two hours in the morning as the stars faded around us. He counted out twenty arm spans of nylon line and dropped it over the side and then he looked deep into the water, waiting, forearm resting on the canoe's edge, the heavy line wrapped three times across his scarred palm. There was always competition, however: younger men motoring along the reef's edge also on the lookout for live fish species. These fishers were not using lines, hooks, or bait—only chalky clouds of poison that erupted out of their plastic squirt bottles, stupefying and taming otherwise wary fish.

Jafir, who only used hand lines, told me that he was torn. He too was caught up in the excitement of the live fish boom, and he wanted to be able to sell fish to the live fish camps near Susunang village. Fish camps, which exported live fish to Hong Kong and Singapore, paid Togean fishers relatively well in local terms. In our conversations, Jafir maneuvered to protect and perpetuate the industry, defending this camp or that as "clean" (*bersih*) and not supporting poison. Yet he also recognized that live fishing had brought a new reality to Susunang—because of cyanide use, there were fewer fish for people to eat.

By the mid-1990s, the live reef food fish trade was proving harmful for the majority of Togean fishers, their communities, and for local coral reef environments. Radiant wild reef fish that had always provided fishers with extra income through centuries-old markets for salt fish, and that

Hand line fishing, by Celia Lowe.

had always been a reliable source of food, were becoming rare in Togean waters. Many government bureaucrats and biodiversity conservationists believed that all Sama people fished with cyanide, and that Sama were the main ethnic group responsible for the damage cyanide inflicted on Togean and other Sulawesi reefs. Once, I was with Yakup when a Sama fisher came to sell us his catch of Fusiliers (*ruma ruma* [S]). Yakup accused the fisher of catching them with cyanide, even though ruma ruma are a variety of schooling fish and cyanide is only ever used to catch a few species of demersal fish like grouper. He had already naturalized an association between Sama people and cyanide-caught fish.

While most live fish were caught with cyanide, not all cyanide fishers were of Sama ethnicity, and the majority of Togean fishers of any ethnicity used traditional hand line techniques. These ratios were disguised within the causes and explanations formulated by concerned international and Indonesian biologists who condemned the live fish trade as inherently destructive, and who blamed Sama people as poor conservators of nature. In 1998 I was contacted by an international conservation NGO that wanted to "reintroduce" hand line fishing to the Togean Islands and encourage Sama people to use this fishing method again. This was the strategy the NGO had used to combat cyanide fishing in the Philippines, and the organization's representative did not believe me when I told him that 85 percent of all live fishing was already done with hand lines in the Togean Islands.

In an interconnected scenario, traders targeted Sama people as fish suppliers, while other outsiders—biologists and government officials—associated Sama fishers in an overly simplified manner with the illegal use of cyanide. The damage done by cyanide fishing is not explained as the independent acts of a few misguided fishers, however. Togean people were caught within the fibers of markets, law, bureaucracy, and identity—all factors determining the patterns of who would fish with cyanide, who would profit the most by it, and who would suffer the consequences.

How should we understand Togean Islands cyanide fishing then, if not as the loss of hand line techniques, or as the consequence of ethnic identity? We must look at markets for live fish, at how traders bound individual fishers to destructive practices that some fishers would rather have been free of, and at the ways that legal frameworks favored business practices over fisher interests, making Togean people vulnerable to police extortion and the enforcement of inconsistent regulations. These different aspects of the live fish trade worked together to make Sama people the victims of degraded environments as well as the assumed perpetrators, while local bureaucrats and outside entrepreneurs became well-off in the process.

The Togean Live Fish Trade

Live reef food fish was one of the periodic boom and bust trades that swept through Indonesia during the Suharto era. At first Togean Sama fishers welcomed the new market. Profits from live fish allowed them to buy cement and wood to improve their homes, and some fishers banked their proceeds in enduring sources of value like watches and gold jewelry. Live fishing brought at least a momentary economic prosperity to the Togean Islands, and Sama people were willing to spend their time at live fishing, with a gambler's eye toward catching "the big one." While live fish were profitable locally, they were even more valuable beyond the Togean Islands. The way Jafir imagined it, one large Napoleon wrasse, a much sought-after fish, was "as valuable as a new car" when sold in Hong Kong. Togean fish buyers, the middle men, were also becoming wealthy from live fish. One fish buyer told Jafir that he would prefer his outboard motor sink to the bottom of the sea rather than have even one Napoleon wrasse escape. Jafir's stories highlighted the economic value of live fish in the Togean economy and in fisher's lives, while pointing out the social and economic disparities between live fish harvesters and live fish buyers, and between suppliers and the people who eat live fish in expensive restaurants overseas. Fish that sold for US$7.50 per kilo in the Togean Islands,

eventually would sell for US$180 per kilo in foreign restaurants (Johannes and Riepen 1995:9).

Despite the initial enthusiasm for the live fish trade, the negative effects of the trade on Togean communities, on fish species, and on Togean coral reefs was of intense concern for many Sama people, for IFABS scientists, and for some government officials. The use of sodium cyanide (NaCN) causes a high mortality rate in the fish and, more importantly, bleaches the surrounding reef, killing coral habitat. More of the reef is ruined when fishers tear apart coral heads to get at the stunned fish hiding in rocky crevices. While Sama people noted general declines in fish abundance, the survival of particular endangered species, especially the Napoleon (Humphead) wrasse (*Cheilinus undulatus*), was of additional concern to biologists. This fish, the most sought-after species in restaurants, very quickly became rare after the introduction of the live fish trade in Indonesia. Marine biologists predicted that live fish species would be commercially extinct in Indonesia, and in some areas locally extirpated, sometime around the year 2000.

Using cyanide does not require much ecological knowledge on the part of fishers. To catch fish with cyanide, a fisher is towed in the water behind an outboard-powered boat and looks from side to side for live grouper and Napoleon wrasse. Once a likely fish is spotted, the fisher chases it into a coral hiding place. The fisher uses a small squirt-bottle filled with a solution of one tablet of cyanide to five liters of sea water. Then while either holding his breath or breathing from a hookah rig attached to an air compressor, he dives and squirts cyanide in each of the holes in the rock. The fish, having nowhere to go, may come out "drunk" (*mabuk*), or else the fisher will tear apart the coral with his hands or a crowbar to retrieve the fish. Once in the fisher's hands, the fish is guided quickly to the surface and put in the boat's water-filled holding compartment.

The Togean Island live fish fishery was only one example of the commodification of wild reef fish that had spread throughout Southeast Asia and much of the western Pacific. According to Robert Johannes and Michael Riepen (1995), the live fish market first developed in the late 1960s with fish stocks coming from reefs near Hong Kong. The trade proved so lucrative that Hong Kong fishers soon began to move further afield. The Philippines became the next main fishing ground in the mid-1970s, and in the 1980s and 1990s live fishing spread into Palau, Papua New Guinea, the Solomon Islands, the Maldives, and Indonesia. By the mid-1990s Indonesia supplied half of the international live fish market.

At first, live fish were caught by foreign fishers who fished illegally in Indonesian waters. The captain arrested in Susunang with his Filipino crew was most likely a remnant of this type of commercial enterprise. Later, the pattern of small collecting companies using live fish transport

TABLE 1
Species Bought in the Togean Island Live Fish Trade

Sama	Indonesian	English	Latin
langkoe'	maming	napoleon wrasse, humphead wrasse	*Cheilinus undulates*
kiapu tikos	kerapu tikus	polkadot grouper, panther fish, barramundi cod	*Cromileptes altivelis*
sunu	sunu/ super	coral trout, leopard grouper	*Plectropomus leopardus*
sunu	sunu/ super	polkadot cod	*Plectropomus areolatus*
sunu	sunu/ super	bar cheeked coral trout	*Plectropomus maculates*
sunurang	sunurang	—	—
sunu macang, kiapu macang	sunu macam	highfin grouper	*Plectropomus oligocanthus*
tembolang	kerapu tembolang	—	—
gomez	kerapu gomez	flowery cod	*Epinephelus fuscoguttatus*
gomez pipi'	kerapu lumpur, kerapu gomez	queensland grouper	*Epinephelus lanceolatus*

vessels and employing local fishers took shape. In the Togean Islands, centrally located fish-buying camps acted as collecting and transfer points for live fish. Camps purchased ten different species of fish, keeping them in pens, feeding and medicating them with antibiotics, and holding them for up to two months awaiting shipment overseas. Live fish transporting vessels then moved the fish to Hong Kong and Singapore. The total annual market for live fish was roughly 25,000 tons in 1995, some of that being cultivated fish, although precise figures were not available (Johannes and Riepen 1995).

Live fish kept in holding pens often died very quickly in Togean fish camps. Mortality was highest for the coral trout (*Plectropomus leopardus*), of which 50 percent of the fish died before transport, and for the grouper species known locally as *tembolang*. Fish camps eventually stopped purchasing tembolang because of their high death rate. Moreover, the mortality rate of coral trout rose twenty percent between 1996 and 1997, and no one knew why. In 1997 the four camps in the eastern Togean Islands were each exporting 200 tons of fish per month, down from 300 tons in 1996. Rising fish prices are one indication of decreasing fish stocks, and prices for all species were on the rise in 1997. Competition

for the fish was also intense at the wholesale level, and new buyers were still entering the Togean market despite the declining fish catch.

The Science of Cyanide

Togean fishers and biologists each mustered different rhetorics to support their empirical evidence, and they employed different aesthetics to bolster their explanatory frameworks. For example, Silvester Pratasik, a marine biologist at Universitas Sam Ratulangi in Manado, exposed fish to sodium cyanide at 1.5 ppm in a laboratory experiment on two freshwater food fish, *Tilapia mossambica* and *Cyprinus carpio*. Pratasik described the symptoms of the poisoned fish, which died within twenty-four hours, in the following way. "[T]he fish breathed on the surface water, jumped out of the water, imbalancedly (sic) swam, put down on the bottom, the fins stood, the body stiffed (sic) and the mouth was widely opened then the fish died and the skin released much mucus" (Pratasik 1983). Sodium cyanide, he explains, works by converting to hydrocyanide (HCN) in the body and absorbing the blood's oxygen.

Pratasik's experiment took place in enclosed tanks where subject fish could not escape to uncontaminated water. In open water fish are exposed to lower concentrations of cyanide, which stun without killing. In these settings, fishers conduct experiments too. Umar described for me the behavior of a cyanide-exposed fish by using his hands to trace its dizzy path. He said, "the fish is drunk, it spins and spins without knowing where it is going." Jafir used metaphors of flow to depict the travels of cyanide through the water column: "Cyanide is like expensive cigarette smoke or perfume; everyone in the room comes under its influence." Or, "it is as though someone hasn't bathed and everyone around has to smell him." Cyanide fishers laughed at the antics of the stunned and disoriented fish and took tactile pleasure in catching wild fish with bare hands. But they did not want to see fish die and they knew how to clear a fish's gills of cyanide to revive it. They had also discovered how to deflate a fish's floatation bladder, which expands when a fish is raised up from the bottom too quickly. Through these methods they could keep fish alive and healthy for sale.

Togean fishers document the ecological effects of the live fish trade by observing fish rarity, changes in fish behavior, coral bleaching, and the systemic effects of cyanide. Scientists and fishers agree that cyanide is not harmful to humans who eat the fish because the poison rapidly works itself out of the fish's system. The availability of cyanide in the Togean Islands has resulted in human mortality, however. Several people have committed suicide by deliberately ingesting the poison. In contrast with

the "experience-far" concerns of scientists, fishers expressed their worries in relation to regular interactions with the coral reef environment. Puah Jafir, for example, demonstrated the effects of cyanide for me by holding up two triggerfish he had caught; one of the fish was fat and the other very thin. The thin one, he said, had encountered cyanide that had made it lethargic. Cyanide causes a fish to lose its appetite, which means that it won't bite a fishhook. Jafir related that he used to catch large Napoleon wrasses right up close to shore, in the open, near to the mangroves, in only one meter of water, right under his boat. "The fish know." he said, "They hear an outboard motor now and they run, they run deep out to sea—they hide from us."

Paradoxically, the live fish trade was generally thought by some biologists to be a sustainable industry that could have widespread positive effects on the incomes of some of the most economically vulnerable fishers. Many Togean Sama fishers knew how to, and did, use hand lines that did not poison reefs and extracted fish at lower, arguably sustainable, rates of harvest. It was not "science" in and of itself that divided Sama from biologists, in this case.

Who Uses Cyanide (and Who Doesn't)

The live fish trade was a multiethnic business made up of people from all levels of society. The role of any given participant in the live fish trade was more closely correlated with class difference than with ethnicity. Live fish businesses were owned by wealthy Indonesians from Jakarta, many of Chinese descent who operated through established connections with foreign Chinese buyers, a pattern having historical roots in other natural resource trades. These owners could afford the capital investments and costs associated with operating a camp. Such costs included fish holding pens, camp buildings, short-wave radios, salaries, and especially permits and bribes. At the local level the owners hired managers to run the fish camps. These managers came from Java and Kalimantan to act as "boss." Bosses fit relatively comfortably into village life and sometimes married local Togean women. At the tail end of this extractive commodity chain, all Togean ethnicities were represented among people who fished. The connection of Sama people to cyanide had come about partly because fish camp managers placed their camps strategically near to Sama villages, and actively recruited Sama participation. Thus, there *were* Sama participants in the trade at all the lowest levels. But Sama were certainly not the only Togean people involved in either live fishing or cyanide use.

The general perception of Sama people as *suku terasing*, a term meaning an ethnic group for whom the process of national development was

still foreign, helped to articulate Sama ethnic identity and environmental damage. Susunang and the other Togean Sama villages were all part of the Program for Left-behind Villages (*Inpres Desa Tertinggal*), a poverty alleviation scheme for putatively backward villages. A consequence of this typology was that Sama people were thought of as inferior to wealthier, urban Indonesian citizens. Sama people were at once ridiculed for their cultural and fiscal impoverishment, and prodded to come up with means for their own development. When some fishers found the means to "develop" through the live fish trade—build new houses, wear new clothes, own motorized transport—they were then chastised for being environmentally destructive.

When it came to the issue of cyanide,. many people in Susunang complained about those using cyanide. For example, on the pongkat, Puah Umar called out to a fisher passing by, "don't bother fishing here, they won't eat your hook. Someone from my village was using cyanide here this morning. You should get your village head to report him." When Puah Padi spotted a person using cyanide she said, "tie him to a rock and dump him in a deep spot," and she frequently insulted cyanide fishers as "rock heads" (*kepala batu*). Ordinary fishers were angry with the ones using cyanide, and families and friends were often divided over its use. Cyanide fishers, they said, are the people who have outboards and are making lots of money, while everyone else's take of reef fish—for both trade and food—disappears.

My experience with Jafir indicates that Sama people might actually have been *less* likely to use cyanide than fishers from other ethnic groups. While cyanide use on coral reefs was frequently attributed to fisher ignorance, Jafir's knowledge of the marine world, in contrast, was quite extensive and directly relevant to the capture of live fish. His hand line fishing, for example, required intricate knowledge of species, currents, locations, equipments, and baits. He told me that Napoleon wrasse will eat all varieties of Squirlfish (*babakal* [S]) and he named eleven separate subspecies of Squirlfish that are appropriate as Napoleon bait. While cyanide fishers swept the seas attempting to harvest every possible fish from every imaginable reef, Jafir sat waiting for a Napoleon wrasse to swim along its favored path.

Non-Sama could sometimes be more inclined to use cyanide if they did not recognize the appropriate baits or spatial tactics to use for Napoleon fishing. As new entrants into the market they were less inclined to take up these difficult practices, and they were more likely to go directly to easier and more destructive ways of fishing. When I interviewed a Javanese migrant fisher about the presence of cyanide-caught live fish in fish camp holding pens, he responded that all Napoleon wrasse are caught using cyanide. Yet, whenever I fished with Jafir and others and observed

them fishing selectively for Napoleon, they employed an ecological knowledge that enabled fish catch without poison. This point has not been lost on some in the Indonesian conservation community where Rili Djohani (1993), for one, has argued that Sama peoples' sea experience could actually make them important marine conservators in Togean and other Indonesian coastal settings.

All Togean ethnicities were represented in cyanide use, and patterns of cyanide practice could be explained more accurately by examining other aspects of identity like age or gender. Since women fishers never use poison to catch live fish, for example, gender was a dividing line separating hand line fishers from cyanide users. One day, when I asked why everyone was trying to catch a small sardine (*solisi* [S]), Puah Padi answered, "we are looking for *solisi* because all the big fish have been poisoned." While to protect their families, women would cover for a spouse or son using cyanide, it was also women—and women fishers in particular—who were the most open in their criticism of the destructive fishing practices they knew to exist. Women's perspectives and participation as fishers and as community members vested in environmental outcomes, were often overlooked.

Nearly all cyanide fishing is carried out by young men. Since it is a physically strenuous activity, older men have difficulty diving and they also complain of the cold. High live fish profits brought by cyanide use enabled young men to build houses and establish new, independent families. Cyanide fishing, as an illegal activity, also had a status that was appealing to younger people. They could earn money to smoke expensive cigarettes and wear fashionable new clothes. The illegality of the technique indicated their closeness to officials, who would protect them from prosecution. While young men caught fish with cyanide, some older men participated by using attachments to bureaucrats they had developed as village leaders. The Indonesian state bureaucracy, which percolated its way down to the village level and radiated out into villages through kinship connections, was tightly correlated with illegal trade in natural resources.

One day a Togean village official, under pressure to appear to be enforcing cyanide laws, ordered sea operations (*operasi laut*), also called sweeping (*sweeping*), and instructed his colleagues to go out and "clean up the ocean." I went along with a party of five sweepers, all wearing khaki uniforms, who set off with their boat drivers in three different directions. Only our boat had any success: we caught five skinny boys, none of them older than ten, using cyanide to catch anemone fish. Children like to play with the clownfish that live symbiotically with sea anemones by making the fish fight each other in small containers of sea water. The boys, all yelling at once, begged us not to report them. The police would be angry,

and angry police are often physically violent. A solution emerged. The boys would deliver edible anemones to the village official's house. Conspicuously, their poison was not confiscated and they were left to resume their activity.

From there we proceeded to a less affluent part of the village, far from where any officials lived, and it was mentioned to the parents that their children had been caught using cyanide. In the same breath the uniformed village official made a casual inquiry as to whether there were any ripe mangoes? He was soon sitting on the porch, chin dripping with mango juice. He asked again for fried sago, which the family procured with ingredients quickly borrowed from a neighbor. "Don't forget the coconut!," he ordered. Coffee with tablespoons of expensive sugar was served after the mangoes; gifts of limes and chilies were taken before leaving. Conversation between the high-status village officials and their subordinate fellow villagers had been smooth, never strained, polite. The threat was always left implied, wound around in-and-out of discussions of mangoes and chilies, sweet and hot.

Fishers involved in cyanide use, if not immediate family members of village officials, were at least closely related to them. Village leaders provided protection against prosecution for their relatives and workers. "He uses a code when he directs them not to use cyanide, which indicates that in his heart he will not really be mad if they do," Jafir explained about one village leader. Ties to bureaucrats were further relevant for determining who paid bribes and who was prosecuted instead. The children collecting anemone fish were from families without strong ties to village leaders and were thus vulnerable to demands for payment. Local leaders of cyanide operations worked intimately with fish camps and tended to channel financial opportunities and protection benefits to family members who they could both trust and control. Outside of this circle, fishers, even small boys, used poison at their own risk.

Subsequent to this "sea operation," Jafir, who opposed cyanide use, complained to me that the manner of our operation was all wrong—not really designed to catch anybody. It was conducted at the wrong time of the day, and not where people really used cyanide to fish. More importantly, he said, the people involved in the operation were themselves all heavily implicated in cyanide fishing. Village level bureaucrats, nested tightly inside live fish procurement networks, worked closely with fish camps that supplied cyanide and bought cyanide-caught fish. Our boat drivers had steady work as cyanide fishers in the employ of these leaders. In other words, we had only been out to catch ourselves.

It would be easy to be swept up by a theory that village leaders who profited from cyanide were just bad people, but these networks originated, and were patterned on, an entrepreneurial culture that starts at the

top of the Indonesian political hierarchy. Neoliberal cultures of economy concur with these resource extraction practices, and powerful Northern nations have acted as guarantors of an Indonesian state that enabled international trade while subverting political opposition. The officially choreographed subversion of opposition in favor of business interests worked efficiently in Indonesia, right down to the village level. Local village leaders were recruited and inducted into this culture by members of the regional bureaucracy, who were in turn invited to participate by national level bureaucrats, and so on. In fact, village leaders would not have been permitted to keep their jobs if they were unwilling to participate in bureaucratic entrepreneurship. Village officials may have reinvented, but surely did not invent, these ways of organizing economic life.

New, Fast, and for Sale

Commodities take on new meanings through their nomadic travels, and live fishing is a cosmopolitan enterprise that stakes claims to locality in many different places (Appadurai 1995 [1986]; Bourdieu 1984; Kopytoff 1995 [1986]). Wild reef fish from some of Indonesia's most inaccessible regions, hearts stopping as they hit a wok or steam pot, are consumed by wealthy elites in the luxury restaurants of Singapore and Hong Kong. Live fish are eaten at celebrations and religious festivals where fish symbolize abundance, and the traditional preference for wild caught, freshly killed fish is based on consumers' beliefs that taste, texture, and healthful properties fade immediately after the fish dies. Where markets for high priced consumer goods tie global suppliers to ever expanding networks of international buyers, live groupers and wrasses caught by subsistence fishers on isolated Indonesian coral reefs fit into stories of the new, the fast, and the "for sale." Markets for live fish, which link Southeast Asia's least affluent people with wealthy global elites, developed hand in hand with rising incomes in urban Asia. Johannes and Reipen have proposed the Hang Seng Index as an indicator for live fish demand—a demand that seems to increase with fish rarity (1995:17).

The link between scarcity and price is a rare gap in the otherwise seamless way the live fish commodity effaces its history. I imagine consumers have little awareness of the genesis of the grouper that arrives at the table in Hong Kong. On either side of the South China Sea, however, the contexts of live fish consumption and live fish production are mutually obscured.[1] Fishers have been motivated to catch live fish by their own economies of luxury through the incentives fish traders give them for their catches. In addition to cash, camps often supplied fishers with gasoline for outboards, cigarettes, and small loans. While consumers would have

a hard time imagining the conditions of live fishing, fishers can only fantasize the extravagant lives of those who consume live fish. Although the sins and seductions of eating live fish are opaque for consumers whose demand motivates live fish production, the social and ecological relations of that production cannot be so easily ignored by fishing communities.

The participation of fishers was even more complicated than willful consumerism. Togean fishers, even those using hand line methods, became tied to particular fish camps through the purchase of outboard motors on credit. Outboard motors are practical in the Togean Islands where sail and paddle are the most prevalent means of moving about, but they are financially out of reach for most people. In the early years of the Togean live fish trade, outboard motors became available to 10 percent of households, largely through loans from fish camps. The camps gave fishers five horsepower motors (*ketinting*) worth US$450 when bought on credit from the camps, US$300 when bought directly on the mainland. Fishers paid for the motors with irregular payments (*uang cicil*) taken out of their live fish sales. The camps maintained the outboards, changing oil and spark plugs, as long as the motor was being paid off. By the time a motor was free and clear of payments, however, the outboard usually was no longer working.

The camps required fishers to pay on their loans each time they sold a fish, but fishers could decide for themselves how much to deposit. One day Narto showed me his accounting books. Narto, who grew chocolate and coconuts as well as trading in live fish and sea cucumber, explained the tie he had to one of the buyers. Every morning he would fish, and in the afternoon he would come home to work in his garden. He was tired of fishing, he said, and would like to spend more time farming. But he had to keep fishing or the camp would take his outboard away. I saw that he made random payments up to four times per month, of amounts between one and six dollars. He could pay off one quarter to one third of his debt in a year, and he was always optimistic about how much he had already paid. Though he wanted to own the motor, he also rued the ties he had to the fish camp.

Likewise, Suala, who liked to fish nearly full time, was using his outboard to fish for pelagic tuna. He needed the outboard to get to the offshore fish concentrators where deep-water fish school. Yet, his camp told him they would confiscate his outboard if he did not resume catching live fish for them. The camp's profits were not made from loans, but from guaranteeing their supply of fish and ensuring that fishers would keep working. So if Suala wanted to prevent the camp from repossessing his motor, he had to continue to fish on a regular basis. As the supply of fish became more scarce, this entailed more of his time and effort.

The relationships fishers had with fish camps were significant for another reason: live fish camps were themselves centrally implicated in the procurement of cyanide. Puah Jafir claimed that poison was not used in the Togean Islands before the arrival of the live fish industry, when fish traders came with cyanide and taught fishers how to use it. The traders then supplied cyanide at no cost. "You shouldn't teach people how to use stuff like this, then it wouldn't be used," he observed ironically. Traders working with the camps also supplied the compressor rigs used to catch live fish in deeper waters and in more difficult locations. Compressors go hand in hand with cyanide use; there is no other way for a diver using a compressor to catch live fish. These techniques have allowed live fishers to target literally every single sizeable grouper and wrasse on a given reef.

Laws from Above, Enforcement from Below

If cyanide fishing was illegal, and many Sama people opposed its use, how did it then come into common practice? The pragmatics of legal practice provide one answer. Along with biomedicine, law is a preeminent form of reason, and an insistence on legal form is one method states use for asserting a universal, unmarked place within the community of nations. In the 1990s Indonesia continually referred to itself as a "legal state" (*Negara Hukum*), and it did have a formal legal framework generous in its protections for both ordinary people and the resources they depended upon. Environmental laws, for example, prohibited the use of destructive technologies, the harvest and export of endangered species, and the penetration of foreign fishing vessels into Indonesian waters. The law is also enacted in practice, however, and Indonesian resource law revealed a consistent support for large business interests and a structure that was suspicious and punitive at the community level. Laws regulated fishers, businesses, and government agencies differently, thereby determining the calculus of gain and loss in the live fish trade.

The basic framework for environmental protection in Indonesia falls under the "Law of Living Natural Resources and their Ecosystems" (GOI 1990).[2] Embedded in a philosophy that nature is God's creation, the law states that the natural environment should be preserved for the common welfare of Indonesian citizens specifically, and for humankind more generally. At a biological level the law acknowledges elements in an ecosystem as interdependent, and ecological processes as part of life-support systems. At a social level it proclaims that ecosystems and species should be utilized in a manner protective of the ecosystem and its flora and fauna. The "Basic Fisheries Act" (Law 9 of 1985, *see* Warren and Elston 1994) is one example of a law focused on popular well-being with aims that

include improving the lives of fishers and preserving the fish resource. Indonesian environmental legislation contains the broad argument that conservation and development are inseparable and both necessary for human welfare; thus, environmental protection has been considered the shared responsibility of citizens and of the state.

Foucault has written, "the bourgeoisie was to leave to itself the illegality of rights: the possibility of getting around its own regulations and its own laws, of ensuring for itself an immense sector of economic circulation by a skillful manipulation of gaps in the law—gaps that were foreseen by its silences, or opened up by *de facto* tolerance (Foucault 1975:87)." Indonesian resource law, which, on the surface, advocates well-being for both humans and nature, ensures a similar "illegality of rights" for an elite segment of the Indonesian population. In a country where "development" has been an ideological imperative (see George 1996; Hefner 1990; Keane 1997; Steedly 1993; Tsing 1993), environmental legislation in Indonesia has created a protected environment for business, while focusing conservation responsibility, enforcement, and blame onto marginalized communities.

Bracketing, momentarily, the argument that Indonesian legislation has never been expected to fulfill populist rhetorics, we can see how official structures have moved from protecting people and ecosystems to protecting the interests of bureaucrats and traders by looking at the legislation for Napoleon wrasse. The decrees, titled "Ban on the Napoleon Wrasse Fish Haul" (GOI 1995a), and "Ban on Export of Napoleon Wrasse Fish" (GOI 1995c),[3] appear in name to insulate this species from catch and sale, since markets were almost wholly foreign and export is banned. The law on export states, "the hauling of the Napoleon wrasse has been conducted using manners that may be harmful to coral reef ecosystems and other marine biology," and, "in the framework of development and conservation of fish resources and coral reef ecosystems, it is deemed necessary to establish a ban on export of Napoleon wrasse fish." Both laws, however, contain the kernels of exception that actually facilitate, rather than hinder, the catch and export of the fish.

Fish camps, exporters, and government officials are the direct beneficiaries of these legal frameworks since fishers are required by law to sell their catch to a "collecting company." Even though in practice these companies control capture methods through the provision of equipment and incentives, the law does not speak to this role and they are hardly regulated or hampered in their ability to sell and profit from the Napoleon wrasse. For example, Article 8 of the Ban on Haul states, "fish shall weigh not less than one kilogram and not more than three," while Article 9 says fish "weighing more than three kilograms or those weighing less than one kilogram will be allowed to be sold locally to a marketing entrepreneur"

(GOI 1995a). Thus, although the law formally disallowed their export, fish weighing too little or too much can legally enter the hands of traders whose only intent is to sell fish abroad. Napoleon wrasse laws also have allowed catch for research purposes, but collecting companies—fish camps—are not set up as research stations. There is not any reason other than export to purchase Napoleon wrasse from fishers, and it is unrealistic to expect that the fish the camps buy will not be exported (or that large fish will shrink to permissible export sizes!).

Laws that enable trade in live fish simultaneously attenuate bureaucracies and enrich individual government workers. This occurs through the government's reporting, evaluating, and permit-granting roles as outlined in live fishing laws. Government agencies grant permits for the haul of fish and require other permits to export live fish. Each Napoleon wrasse also needs an official "letter of origin" for legal export. The provincial fisheries department is further obliged to "determine the fishing ground by evaluating the resource and its environment." Despite this "oversight," the Napoleon wrasse resource was well on its way to local extirpation in the Togean Islands in the 1990s, indicating a deficit in will, funding, and expertise—and even intent—of the state to carry out its role as resource guarantor. This was because the fees collected for permits and services ensured that the government's own oversight practices created conditions for maximum exploitation and minimal protection of live fish and other natural resources.

Every government agency has its agents. In Indonesia, bureaucracies need to secure their own funding for all but the most rudimentary tasks, and the personal income of government workers is directly connected to the "fees" generated through bureaucracy. Granting permits and facilitating trade are the ordinary profit-making activities of many branches of government. Permits for fish camps in the Togean Islands reportedly cost US$1,000 in "official money," implying that immeasurable hidden charges surpassed this figure. It was in the self interest of bureaucrats to grant these permits, not to restrict access to natural resources. Further, exporters without proper permits have illegally identified Napoleon wrasse as "grouper" on customs forms and have paid officials not to inspect their shipments (Johannes and Riepen 1995:40). In Indonesia, permit requirements providing personal income for officials was the rule, not the exception, and there was no evidence to support the idea that enforcement limited, or was even intended to limit, cyanide use.

Unlike traders and bureaucrats, who were situated to profit from the industry without liability, risk, or blame, fishers were vulnerable to prosecution and extortion, and they also had to live with the material consequences of damaged coral reef environments. In another regulation, the "Decree of the Director General of Fisheries Regarding Size, Location,

and Manners of Hauling Napoleon Wrasse Fish" (GOI 1995b), we can see the role fishers are assigned in the interstices between traders and bureaucrats. This law allows for catch and trade by "traditional fishers," defined as persons or groups whose means of livelihood is catching fish using nonmotorized craft or small outboards and, "which utilize fish catching devices and substances that shall not harm the fish resource or its environment." Different types of rules and responsibilities within the decree apply to fishers, businesses, and government agencies. Rules pertaining to fishers focus on techniques and equipment: fishers can use lines, traps, and nets to catch Napoleon wrasse. Emphasis on catch method implicates fishers as the party responsible for how fish are caught. Unlike fishers, collecting companies are attributed "scientific expertise." [4] Yet, since cyanide is distributed by fish camps to fishers, the law implicates only the most vulnerable party. Moreover, fishers' vulnerability is located in their inability to pay the fees and fines, which facilitates extralegal economic activities.

A story of the enforcement of a law prohibiting the use of air compressors to catch fish reveals the manner in which the state and its laws functioned in the Togean Islands at the time. Representatives from the police, Navy, and fisheries departments descended on Susunang village one day to perform what they called a "secret operation" (*operasi rahasia*). They were checking for the permits that are required to own a compressor. One village leader was told to notify the owners of the compressors of their permit violations. Indicating his own involvement in illegality, he answered that he had "influenza" and couldn't leave his house. Unlike the sea operation run by the low-level village bureaucrats, these outsiders were successful in finding culprits, however, and three compressors were temporarily confiscated.[5] The rumor that the equipment owners would be taken to the regency capital, Poso, to face charges traveled around Susunang. The way out of the difficulty involved cash, and the sooner the problem was dealt with, the less expensive it would be.

Togean people consistently expressed the idea that only poor people end up in trouble with the law; those with means can pay their way out of difficulties. As Jafir said, "I want to help [those arrested], but to help with money—there isn't any money. To help with advice—I don't want to seem like I go along with the position of the police. What do I do?" Most Susunang people had $50 or $100 in savings at the most, and release from jail cost the impossible sum of US$5,000, I was told. This forced Sama and other Togean people who were caught in enforcement webs to turn to village officials and entrepreneurs who would trade immediate cash and protection for future illegal resource harvests. Ironically, while poor people were the first to suffer penalties they also assumed the greatest risks, yet were excluded from the highest live fish profits. Laws, as

they were written, interpreted, and enforced within the entrepreneurial Indonesian bureaucracy, enriched bureaucrats and their organizations and failed to protect either species or citizens.

Law as Natural Resource

Two weeks after the initial incident, the story of the arrest of the cyanide fishers from North Sulawesi continued. Susunang's leaders were summoned to speak with a policeman who arrived to investigate the incident. A very imposing figure, the policeman wore a sidearm over his shirt and a chain of bullets lined up across his chest. Rather than supporting Susunang peoples' defense of their local waters, the policeman chastised them for their action. He said that the village had been wrong to detain the boat, and the captain might have to be compensated for his lost revenues with village funds. Susunang's leaders protested that what they did wasn't really an "arrest"—it was simply a chance to have a "conversation" with the boat captain. In the end the village leaders were warned, the policeman was paid some money for the problem to go away, and the captain was free to use cyanide where and when he wished. "The village has the right to regulate its own affairs, but sometimes the police want to mix in," Udin grumbled.

At Uemata, Hari argued that one of the most important steps needed to protect Indonesia's coral reefs is for people to be able to defend their territories from outsiders, who use harmful methods to meet short-term interests. Hari and Laksmi attempted to educate Susunang people on their legal rights in relation to environmental protection. Using the law as a form of community defense was a formidable task, however, given the lack of voice Togean people possess in relation to the Indonesian state. Even when communities are legally right, they are wrong within the practical reason of Indonesian law enforcement. The live fish trade in Indonesia could have been used to benefit a large number of Indonesia's fishers over the long haul, but instead it was designed to make a small number of well-connected officials and entrepreneurs wealthy. This came at the expense of Togean coral reefs and the people who depend upon them.

Pheng Cheah has asked us to consider, "What does it mean for a country in the South to practice hospitality without reserve as a host for transnational capital?" (Cheah 2003:393). The Togean live fish trade provides one answer. Activists and others working for social change in Indonesia were invited to remediate at the community level, broadcasting development and governmentality, and supporting the state's agenda for social control. Some, like Hari, looked for opportunities to advocate for rights. But had IFABS attempted to intervene in natural resource markets, the

state would have viewed this as threatening. IFABS scientists who wanted to assist Togean people to conserve Togean natures were, thus, interpolated into the state's plans for "people's self-help" and forcefully discouraged from interfering in the state's business ventures.

Cyanide fishing is a cosmopolitan practice, however, that must be understood in a variety of ways that take us well beyond fishing communities. Rather than condemn the industry and its fishers, I have tried to illustrate who participates in the destructive aspects of live fishing (occasionally including, but not limited to, Sama people), what structures facilitate their participation, and through what cultural logics these structures have come into being. This specificity is incompatible with liberal theories of the law, however, which are based on the premise that all members of a nation have equal access to its universal capacities, orders, and responsibilities. Laws are based on principles of the good and the fair deemed obvious to "any rational person." It is neither necessary, nor even desirable, from the perspective of liberal legal theory, to understand contingency or social circumstance. Yet the exemplary "human" of a universalist approach to live fish law is also a particular figure: the live fish trader imbued with both knowledge and virtue. It is not Puah Umar or Puah Jafir for whom Indonesian law is not a natural resource.

Chapter Six

THE SLEEP OF REASON

> The sleep of reason produces monsters, but the sleep of reason does not represent the imagination's taking over. On the contrary, reason is asleep when the imagination deserts it. For if, as a contemporary description of [Goya's] work puts it, 'imagination forsaken by reason begets impossible monsters; united with her, she is the mother of the arts and the source of her wonders,' exactly the same is true of reason itself: their relation is symmetrical. It is only when both work together that each is fully awake.
>
> —Alexander Nehamas, "The Sleep of Reason Produces Monsters"

ON OCTOBER 19, 2004, after more than a decade of species inventory, ecotourism initiatives, and social surveys, the efforts of the IFABS scientists finally bore fruit and the Minister of Forestry issued a decree (GOI 2004a) establishing the Togean Islands National Park (*Taman Nasional Kepulauan Togean*). Letters from the governor of Central Sulawesi and the Bupati of Tojo Una-Una Regency supported the federal-level legislation. In justifying the creation of the park, the decree mentions the Togean Islands' natural diversity, including 262 species of corals, 596 species of fish, 555 types of mollusks, pilot whales, and a profusion of land fauna including deer, lizards, the bear cuscus, tarsiers, and the Togean monkey *Macaca togeanus*. The archipelago is further praised for its great potential as a site for ecotourism.

The emergence of the Togean National Park after more than a decade of conservation work provides an ending to my story of the *Wild Profusion*, but certainly it is also a new beginning, a change that will allow new conditions of negotiation and meaning for all involved. It is unclear how corporate pearl farms, live fish camps, or logging corporations, for instance, will fit into this new organization of natural resource space. Will the existence of the park rationalize natural resource harvests in such a way as to shelter the islands and their peoples from the next extractive boom cycle in terrestrial or marine products? What will happen when the live fish trade expires? Will it be replaced with a trade in ornamental live fish as predicted, and will the park make a difference in how this trade is organized? Some Togean people will inevitably ally themselves with the new park, and others will certainly oppose it. New identities and new

natures will emerge, and different social forms and rationalities will be assembled as new natural and national cultures of nature come together.

That biodiverse nature could be produced from Togean Island materials in the 1990s depended upon a series of arbitrary and contingent events. Colonial natural historians, avoiding the mainland due to threats of violence and piracy, established a scientific record that would lead contemporary scientists to the archipelago. A Swiss man named Sibley released a group of macaques in the Togeans in the 1930s, attracting Dr. Supriatna to the islands, and leading him to build Camp Uemata and establish the Indonesian Foundation for the Advancement of Biological Science. There is little doubt that "biodiversity" would have emerged from a decade of taxonomic study in almost any nonurban region of Indonesia, and nearly all of the species listed in the park decree are found in other places. It is also clear that the Togean macaque helped legitimate the park without the support of scientific results. Yet, when the work of Togean conservation biology was completed, other "allies and resources" were made redundant (Latour 1987:98), and the Togean Island National Park now appears to be based on the fact of nature itself.

How certain aspects of the physical world become important at particular moments in time cannot be understood entirely through rubrics of the natural or universal, I have argued. Positing biodiverse nature as prior to signification or representation distorts the effects of precisely what scientists choose to notice, and why and how this attention is constitutive of nature as an object.[1] The assertion of a prediscursive sphere of nature should be understood, rather, as a type of political claim. In the case of the Togean Island National Park, claims to a fundamental, prerepresentational space of nature were a form of instrumental reason facilitating intervention on behalf of plants and animals that cannot "speak for themselves" (Latour 1987). Through conservation biology, Indonesian scientists have been able to claim this transnational universalist position for their own. But to the extent they have aligned themselves with international norms for biodiversity conservation, they have forwarded an argument against Togean peoples' natures.

The formalization of the park provides an opening, however, to organize conservation around Togean peoples' natures (as the IFABS scientists were able to do at some points) rather than scientists' biodiversity. To proceed in this way would mean to recognize how plants and animals have informed cosmopolitan Togean worlds for at least several centuries. It would also mean to problem-solve around the mutual desires scientists and Sama people have for species abundance and diversity. Clams, for example, or grouper, or sea cucumber would all be excellent locations from which to begin to think through a collaborative idea of species conservation. Producing Sama people as a different kind of "indigenous"

human would not be as useful to such work as recognizing them in projects of friendship, thought, or care. To organize conservation around Togean people's natures would mean to refuse reason's affiliation with historicist thought, to disavow tribunals of reason, and to recognize that objects are not all that is needed to know or to think. To organize conservation around Sama people's natures would mean for scientists to begin to take nature at least as seriously as Sama people themselves do.

Biodiversity Conservation Becomes a New Form

I have described Togean conservation as an assemblage of science, nature, and nation in 1990s Indonesia that took into account: the exponential reduction of species forms across the globe; the rise of Southern expertise in transnational conservation projects; the ICDP as a means of reconciling the relationship of certain peoples to plants and animals; and the domestic political milieu of the late-Suharto era in Indonesia. As such, the Togean project represented a particular temporal convergence. With Suharto's demise in 1998 the scene would shift: the changing political field in Indonesia would come into play with the new ecoregion concept in biodiversity science, loss of confidence in the ICDP, and a transformed post-9/11 international political climate. Biodiversity conservation in Indonesia would become a new form.

The fall of Suharto inaugurated the era known in Indonesia as "reformation" (*reformasi*). In addition to anticorruption campaigns, reformasi brought about efforts at administrative decentralization and attempts to address long-standing debates over Federalism left unresolved since the Indonesian Revolution. New "regional autonomy" laws (*U.U. Otonomi Daerah*) gave increased powers to the provinces (*Propinsi*) and regencies (*Kabupaten*) to manage their own affairs (GOI 2004b, 1999a). These laws developed a new system of benefit-sharing between the regions and the central government, and have allowed for changes such as local control over the coastal sea within a twelve-mile limit. Law No. 31/1999 Concerning Fisheries (GOI 1999b) specifically passed authority over coastal resources to the regencies.

At around the same time the major international environmental NGOs began to argue that the ICDP model had failed, and that their organizations were not capable of being everything to everyone. Caring about people by providing economic development or health care draws resources away from the more fundamental mission of preserving species and habitats.[2] Through this reasoning a transformation has occurred in how the international groups problematize the relationship between nature and the human. Reversing Foucault's well-known formulation, the

"human" who threatens nature is now no longer someone to be aided, a figure known through the disciplinary power of governmentality, but a threatening figure who must be controlled through law, surveillance, and sovereign power.

An unintended consequence of the shift from centralized authority to local autonomy in Indonesia has been that the provinces have become more amenable to direct influence. In this context foreign conservation NGOs have used their capacity to fund, train, and equip local constabularies and judiciaries. CI, TNC, WWF, and USAID's Natural Resources Management Project have all become at least peripherally affiliated with, and sometimes directly involved in, projects of "enforcement." These projects have been aimed, for the most part, at the local level rather than at markets or traders. The new concern with enforcement is revealed in a report by CI's "Center for Conservation and Government," which provides "the first quantitative evidence of exactly how poor enforcement in biodiversity-rich countries is":

> Using a behavioral economics framework called the "enforcement economics model," this study quantifies disincentives generated by enforcement regimes and compares them to the profits that motivate large-scale commercial illegal activity in the hotspots. The enforcement disincentive is determined not only by the value of penalties; but also by how likely lawbreakers are to be detected, arrested, prosecuted and convicted so that penalty is incurred, and by how long the system takes to work. The results of this analysis of risks vs. rewards clearly demonstrate that weak enforcement regimes are generating a shockingly insufficient deterrent to illegal activity. (CI 2004:1)

Through an example from the Philippines, the CI study further explains, "in the Philippines' Calamianes Islands, fishermen practicing illegal dynamite and cyanide fishing risk only nine cents, but stand to earn an average of $70.57 per trip." The ethnographic complexities of cyanide fishing that I describe in chapter 5 seep away in the hyperrationalized calculations of the "enforcement economics model." In the Togean Islands, while a hypothesized "average" person might risk only nine cents, real fishers risk $5,000 to get out of jail, or their independence as they become reliant on village officials, or even their bodies and lives should they be caught and beaten as sometimes happens. Many also comprehend the ecological risks they are taking. While the measurement "nine cents" serves as a rationalization for enforcement, it tells us little we do not already know about live fishing, and yet this type of calculus persuades the international groups that they must now turn their energies to developing and enforcing natural resource laws.[3]

As they have shifted attention away from livelihoods and toward militarized and legal interventions, the ICDP has begun to seem anachronistic

and misguided (Lowe 2003). A former representative of USAID's Natural Resources Management Project describes his perception of the shift this way: "If one examines enforcement programs around the world that are fighting other blatantly illegal (but economically enticing) activities that are deemed harmful for the future of society (narcotics peddling, child pornography, and even hired murder come to mind), very few indeed seem to have 'alternative livelihood' programs attached to them. Psychological counseling, perhaps, but not extensive and expensive efforts to retrain drug sellers to become grocery story owners" (Erdmann 2002).

Within this new context the international conservation community has begun to train and equip the police and navy, and they have also encouraged communities to patrol and police themselves. Additionally, they have begun to support the Indonesian judiciary aiding in the process of gathering evidence and prosecuting offenders. The international organizations claim some success, yet there has already been one case brought before the Indonesian Commission for Human Rights (*Komnasham*) when two men accused of illegal fishing in Komodo National Park were shot dead by forces supported by The Nature Conservancy (World Rainforest Movement 2004, Walhi 2003a and b).

An argument can be made about the international milieu within which this turn to enforcement has occurred. George Lakoff (2004, 2002, 1995) has analyzed the current conservative U.S. political climate as the triumph of "strict father morality" over "nurturant morality." In his strict father model, "Life is seen as fundamentally difficult and the world as fundamentally dangerous. Evil is conceptualized as a force in the world, and it is the father's job to support his family and protect it from evils—both external and internal." (1995:10) Both "strict father" and "nurturant" political inclinations have always existed in American society, he argues, and each of us, as individuals, employ dual paradigms. In the post-9/11 world, however, the frame through which Americans view politics has shifted, on balance, in the direction of strict paternalism. With the "global war on terror" this U.S. political momentum has been internationalized.

The permeability of the Indonesian provinces to conservation capital coincides with the transnational milieu of forceful, direct intervention in the "war on terror." Within this milieu, U.S.-based environmental NGOs imagine enforcement as a way to protect nature from the absolute, universally self-evident evil of biodiversity loss. What are the implications for Indonesia of hosting this powerful paradigm promoted by wealthy international conservation groups? Enforcement efforts and projects to rationalize Indonesian resource law are not attempts to create the conditions for "justice" in Indonesia—they do not, for example, include the means to balance a prosecution with a defense. Nor do they represent the adversarial legal system through which environmental conflicts are mediated

in the United States. They are interventions in the name of sovereign power in a country where some one million people died as a consequence of state violence under the Suharto regime. The rationality of enforcement is incapable of taking into account the specific histories or implications of this violent legacy.

Within this new assemblage Indonesian scientists continue to occupy a position in the middle, between, on the one hand, transnational rhetorics of enforcement and, on the other, their own understandings of Indonesian history, politics, and citizenship. In recent work in Susunang village, a "sentry house" was established with village and CI funds to guard part of the reef (Ibid.). At the same time, through a decade of developing friendships and loyalties to Susunang people, some of the scientists have been able to mediate between the universalist ambitions of transnational biodiversity conservation and the particular affiliations they have formed with Susunang people. One scientist describes his view of the new situation in Susunang like this:

> The majority of Kabalutan [Susunang] people are from Bajau or Sama ethnic groups who are economically and culturally bound to marine resources. Unfortunately, there has been a widespread negative perception of them. Most people in Togean Islands refer to Bajau people or Kabalutan people as the perpetrators of fish intoxication or bombing. Whereas in fact, there are larger numbers of Bajau people who want to put an end to those activities, as conveyed personally by the people of Kabalutan village. They have come to realize that their catch is diminishing. Some of them even feel uncomfortable to know that Bajau people are labeled as destroyers of coral reefs. (Sundjaya 2005)

Across Indonesia scientists and environmental activists have responded in a variety of ways to the new opportunities opening up in the post–Suharto period. A more robust Indonesian NGO movement has developed, and these NGOs desire to ameliorate the inequities of the Suharto era: they investigate land tenure claims on behalf of farmers, work on community forestry projects, monitor foreign mining operations, facilitate between communities and bureaucrats, develop "indigenous peoples" alliances, start political "green" parties. Some IFABS scientists have made an investment in understanding social complexity by leaving the field of biology to pursue advanced study in anthropology, geography, or rural sociology. Others have taken advantage of the political opening to advocate on behalf of Togean Island peoples in new ways. Two new groups, the "Alliance of Indigenous Togean Peoples," and "Togean Women's Solidarity," formed in 2001 to protest logging and government land appropriation in the archipelago (Faisal 2001).

Within these projects, new political imaginations of possible "natures," figures of the "human," and relations between the two proliferate and influence the consciousness of Indonesian biologists who continue to work in the Togean Islands. These imaginations are not always harmonious. Some activists accuse the Togean project of "CI-ization" (*CI-isasi*), while other scientists are said to be "too romantic" (*terlalu romantis*) in their support of "indigenous" or "*adat*" peoples (although no one is ever said to be too romantic about science!). An emphasis on forms brings our attention to the historically situated conditions of possibility within which such debates occur. To track the possibilities and limitations of the new *reformasi*-era form of Indonesians' environmental and social activism in greater detail would be exciting work.

Reason and Enchantment

As Alexander Nehamas argues in relation to Francisco Goya's 1799 painting, *El Sueño de la Razon Produce Monstruos*, reason and imagination are both necessary for thought to be fully "awake." It seems that ghosts must be banished from narratives of biodiversity, however, even though their presence is palpable within conservation biology. Relations between universal reason and enchanting myth in postcolonial Indonesia are reflected in the state's formal approval of five "world religions" (Islam, Christianity [Protestantism], Catholicism, Hinduism, and Buddhism) and rejection of what it calls "animism" (*animisme*). This official position on haunting is important for understanding the value of scientific rationality in Indonesia, although it tells us little of the active social life of religion, of Indonesian spirits, or of any phenomenon not otherwise amenable to scientific observation. Max Weber's famous dictum that instrumental reason disenchants the world, creating therein an "iron cage" (what Foucault has called a "*monstre froid*"), is equally applicable in Indonesia where reason's (imagined) triumph over enchantment has meant that the spirit world itself has become inhabited by the cold monster of governmental rationality. Compulsory de-magification haunts the postcolonial nation and the stories it can tell about itself.

It is not only the postcolonial world that is inhabited by phenomena not easily explained through scientific rationality, I have argued. The traffic between reason and enchantment is an important part of any story of science-in-the-making. Johannes Fabian, for example, has described how colonial science was conducted within the context of reason's alter ego: madness. Fabian studied the "uncontrolled, ecstatic aspects of ethnographic knowledge production" in the African explorations of German colonial scientists in the nineteenth century (2000:13): "A truly radical

Order ID: 104-9754492-9576234

Thank you for buying from Three Monkeys on Amazon Marketplace.

Shipping Address:		
Saya Lindsay	Order Date:	Sep 3, 2009
TRINITY COLLEGE	Shipping Service:	Standard
300 SUMMIT ST	Buyer Name:	Saya Lindsay
HARTFORD, CT 06106-3100	Seller Name:	Three Monkeys

Quantity	Product Details
1	**Wild Profusion: Biodiversity Conservation in an Indonesian Archipelago (In...** **Merchant SKU:** GR-G9FV-1ECR **ASIN:** 0691124620 **Listing ID:** 0404E9PW2PK **Order-Item ID:** 09985244039914 **Condition:** New

Thanks for buying on Amazon Marketplace. To provide feedback for the seller please visit www.amazon.com/feedback. To contact the seller, please visit Amazon.com and click on "Your Account" at the top of any page. In Your Account, go to the "Orders" section and click on the link "Leave seller feedback". Select the order or click on the "View Order" button. Click on the "seller profile" under the appropriate product. On the lower right side of the page under "Seller Help", click on "Contact this seller".

critique needs to address the very concept of rationality, especially the built-in tendency of that concept to present itself as outside and above historical contexts." Rather than a pure space of disenchantment, Fabian found the ecstatic, the drug-induced, the magical, the erotic, and the emotional, all relevant to the practice of making knowledge.

Likewise, I view as magical a relationship between the enumeration of species and the ability to conserve nature's creatures in the context where biologists will nearly always claim that "social knowledge" is more important to conservation than the science of biology itself. This is not meant to take away from the rationality of Indonesian scientists at this late date—if anything, they share the magical or mad aspects of science in common with German colonial explorers and EuroAmerican biologists. Rather, the enthusiasm for enumerating and classifying species in transnational projects of biodiversity involves both a haunting by the ghost of natural history, when such a technics of nature was first enlivened, and an act of conjuring where we cannot determine what goes on out of sight when preservation gets pulled out of the hat of enumeration and classification.

While conservation biology normalizes nature as "biodiverse," it simultaneously naturalizes both the human of universal posterity for whom nature will be saved, and that particular human figure who threatens nature (expanding third world populations). While the being who threatens nature has been extensively elaborated in projects of conservation around the world (thus, the emphasis on "development" or "enforcement"), the human for whom nature will be saved has been left relatively unarticulated. Like live fish regulations, where the fish trader stands in for the citizen able to assume rights and responsibilities, or as in biomedicine, where the patient with access to dense allopathic networks possesses the universal human body, the "ecotourist" is the specific figure for whom nature will be saved. Despite claims to the universal value of biodiverse nature, this is a nature only some will be able to avail themselves of, while others should be kept apart from it.

To study science in this way is not to deny the materiality of nature, nor is it to claim that science is not "true." It is simply to make note of our social investments in the scientific questions we ask. Likewise, we can bring our attention to the questions we ask as anthropologists or science studiers. One of the most important recent approaches to the conjuncture of people and nature has connected the environment to questions of social justice. Through the rubrics of "political ecology" and "environmental justice," scholars have excavated the unequal distribution across populations of environmental benefits and risks using an analytic of political economy. It is clear from these studies that there are always winners and losers when it comes to questions of defining and distributing nature, and

these approaches have been crucial for reorienting a generation of thinkers from conservation biologies toward the greater emancipatory possibilities of "liberation ecologies" (Peet and Watts 1996).

From the perspective of an analytic of reason, however, one intrinsically rationalist paradigm does not necessarily free us from another, and an engaged and rigorously articulated critique might not need to divide the world between good and evil. Holding onto a sense of a priori fairness is not necessarily the only, nor even the most apposite, antidote to universalist perspectives on nature. We are all susceptible to the complexities of partial understandings, to fantasies as well as facts; this is the lesson of "studying up." In this book I have approached the problem space of nature and the human through an analytic of reason, not because we can ignore critical questions of social justice, but rather for the possibilities reason presents for conceiving of the issues in a new way.

Reason opens biodiversity's "universality" to reflection and presents its boundaries and possibilities as unstable. While liberal theories of justice suggest that there are clear winners and losers (or domination and resistance), reason contains a more oblique sense of the possibilities for remedy: all any of us have to work with are corrupted forms. In thinking through questions of Togean nature, Indonesian scientists and Togean people are not self-knowing "individuals" possessed of right, nor are they even necessarily free from coercion or physical domination (this is a different thing, and perhaps even more important). Rather, they assemble matter, language, and technique in new ways within already existing degrees of freedom and constraint.

To search for alternative political imaginations in the Togean biodiversity project has meant, therefore, analyzing how Togean peoples' and scientists' "lines of flight rapidly get recuperated, organized, systematized, programmatized," just as we have witnessed these same Indonesians, "creating something new within the most traditional political forms" (Rose 1999:280). We should not expect to find alternative political futures within purified spaces of enlightened or reasoned fairness. This politics is more complex, contingent, and compromised; it come in "degrees"; we might not find it where we expect it, or even know it when we see it.[4]

Likewise, I have described what it would mean to understand the deep imbrication of Sama lives and encompassing Togean land and marinescapes without imagining this relationship as essential in either rationalist or Romantic terms. I have tried to avoid the temptation to read nature and identity as a determining relationship, for example in the figure of the "subsistence resource user," or as a carceral relationship, for example in the figure of the "local person." To explore reason from Togean Island examples has meant to ask what materials are assembled by

Sama people to produce what order of things. To follow the forms of Sama people's nature-making has meant tracking emergent rationalities and practices of thought rather than codes found in the mind.

Independence Day

In 2001 I traveled for the first time to the Philippine city of Dumaguete in the Visayan archipelago. On Negros Oriental and the other Visayan Islands one can find Sama people living along the coast there. Over the past decades they have been driven out of the islands of the Sulu sea by fighting between the Philippine government and the Moro National Liberation Front. Since the 1970s their villages have been bombed, seaweed farmers from other Islands have taken over their fishing grounds and, with the breakdown of traditional patron-client relationships with the Tausug, many have been forced to flee north. In Dumaguete I met one of these refugees, named Boi.

Boi did not have a village full of family and friends he had known all his life—he had lost his community when he fled Sulu. Now he moved between patches of available beachfront and squatted in abandoned house lots with his wife and two grandchildren. Their house was made from blue plastic tarps tied to a platform of driftwood they had gathered. They did not have any access to rattan, bamboo, nipa, or sago palm from the land to construct the beautiful houses Puah Jafir or Mbo Poteiang had in Susunang. Boi's boat, likewise, was not Mbo Dinda's elegant *soppe* or Puah Umar's *leppa* made from single forest logs. He had stitched together a canoe from scraps of plywood using old rags and glue to caulk the seams. Nor did he have a *gonggang* from which to harvest sago and, with no Graveyard Island, where would Boi be buried when he died? Sama people in Dumaguete were known as beggars who were forced to solicit alms in the central market. Through charity and his little bit of fishing, Boi procured the necessities of bare life.

I fantasize Boi's prior existence through H. Arlo Nimmo's lyrical account of Sulu in the 1960s (Nimmo 1994), and I am haunted to this day by the apocalyptic Sama futures I witnessed in Dumaguete. Knowing alternative Sama lives, my heart was heavy as I left the city behind. Boi's existence is what it means to be unable to harvest the land or sea products that have sustained Sama lives for millennia. It represents what it is to truly become a "sea nomad," to have no rights to the flora or fauna of the land, and to lack the familiar sustaining relationship people in Susunang have with the sea. This is the nightmare scenario of a transnational conservation able to fully enforce each of its species prohibitions.

This is the endgame of a global biodiversity vision where elite people on one side of a hemispheric divide are able to separate those on the other side from a nature they have claimed as biodiverse. This is reason deserted by imagination.[5]

But Dumaguete is not the only possible Sama future. On August 17, 1996, I participated in Susunang's observance of Indonesian Independence Day, which took place on the volleyball field in front of the elementary school. At first I misunderstood the event. Having watched the flag-raising ceremony, and having stood in the hot sun listening to overly long official speeches, I got up to go home. I had left the ceremony in frustration, regretting, as I imagined it at the time, that my friends in Susunang were being made fools of (*kasih bodoh*) by a state that believed they were alien to it. But later in the day Puah Hamid asked me why I hadn't stayed until the end. Hadn't I liked the way they had performed the ritual? Didn't they look smart in their khaki uniforms?

I was ashamed that he noticed my absence and that I didn't have the patience to sit attentively. What I did not comprehend at the time was that, not only was the state making an instrumental claim on Puah Hamid, Puah Hamid was also making an affective claim on the nation. While he could not reject the state's compulsory forms, Hamid could assert his membership in the cosmopolitan Indonesian community that claimed difference at the scale of the nation. His was a claim to national belonging for Sama people whose difference seemed to exclude that possibility. He asserted this claim not as an extraterrestrial other, but in the nationally recognizable language of Indonesian citizenship. It was an argument for recognition as subject in, not object of, the national polity.

It is no coincidence that the nation, itself haunted by the instrumentality of the state, operates at a scale that, by definition, resists universal reason. Distinct from the subaltern studies scholars who regard the nation as a reactionary form for its roots in German Romantic thought, its connections to ethnic fundamentalism, and its subordination of the individual to the collective, Pheng Cheah argues from Southeast Asian materials that the postcolonial nation is "the most apposite figure for freedom today" (Cheah 2003:395).[6] I find an argument for the postcolonial nation as a libratory figure useful for taking seriously Indonesians' evident commitment to the nation-form. The "difference" of Indonesians' science lies in a belief in the radical potential of their nation—a nation that has successfully cleared a space for Indonesian scientific expertise—and in Indonesian biologists' faith that they share with Sama others this hopeful imaginary of libratory national futures.

When E. E. Evans-Pritchard wished to demonstrate the common humanity of Europeans and Africans, he was forced by his society's own self-representation to demonstrate a "Nuer" or "Azande" ability to reason.

Likewise, cognitive anthropologists, since the mid-twentieth century, have proven that to classify is human—all peoples have this reasoning capacity even when they categorize differently. To frame reason, instead, as a political claim to knowledge at a universal scale, as a folk category that particular groups of people invest heavily in at particular moments, and as a metacommentary on what will be noticed in a given place or time frees us from this structure as the only means to heal hemispheric divides. Reason, lacking imagination, produces monsters, not humanity. To imagine, to dream and not know that one is dreaming, to be haunted, to be amenable to mystification and ideology, to experience care or friendship, these are also necessary aspects for thinking through the value and meaning of the wild profusion, the nature of the human, and for imagining their common future together.

Appendix

Scientific, Military, and Commercial Explorations in the Togean Islands and Vicinity: 1680–1999

1682*

Jan Van der Wal, captain of the Dutch vessel *Brandgans*.

On behalf of the Governor of Ternate (Robertus Padt-Brugge), Van der Wal produced the first hydrographic chart of the Gulf of Tomini including Togean Islands, and wrote extensive notes on his encounters with Togean peoples.

1855

Edward B. Hussey Jr., captain of the American whaling vessel *Peruvian*.

Described the north coast of the Togean Islands and the presence of whales in the vicinity. Did not sight either people or communities.

1864, June 21–23

Carl Benjamin Hermann Von Rosenburg, scientific officer of the Netherlands Indies.

For two years, including three days in the Togean Islands, studied the geography, geology, flora, fauna, and ethnology of the Gulf of Tomini. Spotted an unusual land snail on Togean Island, and commented on the abundance of sea cucumber for trade, but otherwise found the archipelago to be without natural historical interest.

1871, August

Adolf Bernhard Meyer (born: Aron Baruch Meyer), Director of the Royal Museum of Zoology and Ethnology at Dresden from 1875 to 1906.

Made collections of the birds of Celebes from July–September in the Gulf of Tomini, and in August in the Togean Islands. Made ethnological notes and collected word lists in local Sulawesi languages. Named new Togean bird species.

1878*

S.C.J.W. van Musschenbroek, Dutch resident of Manado from 1875–76, hobby naturalist, and first director of the Colonial Museum in Leiden.

Official responsibilities took him to the Gulf of Tomini, where he produced a hydrographic chart, and a series of explanatory notes from secondary sources on local ethnology and natural history.

*Based on date of publication mentioning scientific exploration; specific dates of travel in Togean Islands or Gulf of Tomini unknown.

1886–1891
Baron G.W.W.C. van Hoevell.
Made seventeen trips around the Gulf of Tomini between March 1, 1886, and April 1891, and described the south coast of Great Togean island. Wrote an explanation of van Musschenbroek's Gulf of Tomini chart based on his own travels aboard the Royal steamships *Java*, *Havik*, *Valk*, *Sperwer*, and *Kamphuijs*.

1900
Nicolaus Adriani, Dutch linguist and delegate of the Netherlands Bible Association.
Studied and recorded the languages of Central Sulawesi. In the Togean Islands, documented Bobonko and Bajau/Sama languages, including a legend of The Monkey and The Turtle.

1928, September
J.H.F. Umbgrove, scientist aboard the *Eriadnus*, a hydrographic vessel of the Dutch East Indies Navy.
Described coastal morphology and inventoried the coral genera of the Togean Islands. Tested "glacial control" and "subsidence" theories of coral reef formation based on his Togean observations. Coined the term "geopoetry."

1939, December
J. J. Menden, naturalist.
Collected macaque specimens from Malenge Island, which he later deposited in the Bogor (Buitenzorg) Museum. Also collected Togean and other Sulawesi bird specimens, which were distributed to museums in Dresden, Berlin, and Amsterdam.

1945, May 3
Lt. Etheridge, U.S. Navy Captain of B-24 fighter plane.
Crash-landed with crew off the village of Pulo Annau, Togean Island, after engine caught fire. Made notes on reactions of Togean people, whom he described as mostly afraid, although a few waved hello.

1976, October
Japan International Cooperation Agency.
Upon request of M. Gobel, Chairman of Indonesian-Japanese Joint Ventures Association and member of Indonesian Parliament, conducted feasibility study for logging the Togean Islands. Produced list of marketable Togean tree species and development plan for their extraction.

1977–1978
Francois Zacot, French anthropologist.

Conducted ethnographic research in Torosiadje, a Sama community on mainland Sulawesi coast directly north of Togean Islands. Began studies of Sama people on Nain Island off Manado in 1976.

1979–1980, December 31–March 31
Victoria Selmier, German biologist.
Conducted study of the deer pig (*Babyrousa babyrussa*) on Pangempan Island near Katupat village; discovered five individuals, including one breeding female. Surveyed deer pig population on other Togean Islands as well.

1982, February 18–25
United Nations Food and Agriculture Organization.
Indonesian and expatriate research team examined marine conservation potential of Togean Islands on behalf of the Indonesian Directorate for Nature Conservation. Concluded the archipelago had high conservation potential for Coconut crab conservation and tourism development.

1987, July 12–October 1
Project Bat, Oxford University Expedition to the Togian Islands (*Proyek Kelelawar*).
British and Indonesian research team surveyed twenty-one cave sites and found sixteen bat species, none new to science. Explained the role of bats as pollinators to Togean people they encountered, and assessed threats to Togean bat populations. Discovered new species of millipede.

1989, November 12–15
World Wildlife Fund/Indonesian Ministry of Forestry.
Indonesian scientific team identified archipelago as excellent location for Coconut crab conservation and tourism. Developed checklist of coral genera from secondary sources.

1991, October 11–29
Deutsche Forest Consult, Germany, and Hasfarm Dian Konsultan, Indonesia.
Indonesian and German consultants studied traditional land rights, marine biology, community development, buffer zone potential, and medicinal plant processing in the Togean Islands on behalf of Government of Indonesia and Asian Development Bank.

1992–present
IFABS/CI.
Indonesian natural and social scientists conducted extensive biological and social surveys of the Togean archipelago. Organized by Jatna Supri-

atna to facilitate primatological research in the Togean archipelago, a field research station named Camp Uemata was built on Malenge Island. Work culminated in the establishment of the Togean Island National Park. A Togean water monitor, a Hawk owl, and other new species discovered.

1993
Sejati Foundation.
Team of Indonesia filmmakers and researchers conducted ethnographic survey of Sama/Bajau communities near Torosiadje on north coast of Gulf of Tomini, for the exhibition *Bajau* held in Jakarta at the end of 1993.

1994, 1995, 1996–1997, 2000
Celia Lowe, American cultural anthropologist.
Lived in Susunang and Camp Uemata for total of twenty-four months. Went bird watching, caught lizards, collected sea cucumber, made sago. Discovered the batunang (*Labidodemas seperianum*) and the sea grape, which were new to her but not to science.

1996, April and August
Earthwatch.
Led by Indonesian scientists, U.S. and Australian volunteers conducted coral reef, fish, bird, and herpetological inventories.

1996
Myron Shekelle, American biological anthropologist.
Studied speciation of Sulawesi tarsiers, including Togean tarsier using genetic and acoustic analyses. Argued for taxonomic separation at the species level of the Togean population of tarsiers.

1998, October–November
Marine Rapid Assessment Program of the Center for Applied Biodiversity Science, CI.
Carried out inventory of marine species, including fishes and corals in Togean and Banggai Islands. New Wrasse fish discovered.

1999, September
Tethyana Expedition.
Australian expedition including Dutch and American scientist. Searched for the origins of the Togean archipelago and its biodiversity. Discovered species not found on adjacent reefs but found in waters of Papua New Guinea. New Acropora coral discovered.

Notes

Introduction: Between the Human and the Wild Profusion

1. I use pseudonyms for all Indonesian scientists and Togean people in this book, with the exception of already prominent scientists who are considered public figures and whose work I am citing. In order to further disguise the identities of both scientists and Sama people, I have also chosen to create composite characters. All of the anecdotes and examples I use are actual things people said or did, but I have arbitrarily combined and separated incidents among my pseudonymous characters.

2. I have chosen to use the nationalized spelling of "Togean" rather than the former colonial spelling of "Togian" or the Dutch name "Schildpad (Turtle) Islands."

3. According to volcanologist Chris Newhall, Una Una Island, or Gunung Colo, has had two or three eruptions. The most recent eruption began on July 18, 1983 and fizzled out by December. A similar eruption occurred in 1898, and a minor event occurred sometime in the late 1930s.

4. The term "BioDiversity" was first coined by the Board of Basic Biology for the "National Forum on BioDiversity," which convened September 21–25, 1986, in Washington, D.C. The Forum was sponsored by the National Research Council's Commission on Life Sciences, and by the Smithsonian Institution's Directorate of International Activities.

5. "Nature" in this book is an empty signifier. In its most general sense, I use it to mean plants, animals, land, and marinescapes. But I also use the word as a placeholder for all of the culturally and situationally specific content that I then describe particular people filling it with. For this reason, I write of "biodiverse" nature, "curious" nature, "Sama" natures, etc.

6. Edward O. Wilson writes, "specialists in this young scientific discipline conduct their studies with the same sense of immediacy as doctors in an emergency ward" (Wilson 1992:228).

7. One of biodiversity's founders and most eloquent and prolific spokespersons, Edward O. Wilson summarizes what is at stake for him in the following terms:

> Biological diversity must be treated more seriously as a global resource, to be indexed, used, and above all preserved. Three circumstances conspire to give this matter an unprecedented urgency. First, exploding human populations are degrading the environment at an accelerating rate, especially in tropical countries. Second, science is discovering new uses for biological diversity in ways that can relieve both human suffering and environmental destruction. Third, much of the diversity is being irreversibly lost through extinction caused by the destruction of natural habitats, again especially in the tropics. Overall, we are locked into a race. We must hurry to acquire the knowledge on which a wise policy of conservation and development can be based for centuries to come. (Wilson 1986:3)

8. Perhaps the most famous example is the position held by Richard Leakey as Director of the Kenya Wildlife Service from 1989 through 1994.

9. Beginning in 1982, multiple excursions by scientists to the Togean Islands documented their potential as a nature reserve. These analyses are described in reports by the UNDP Food and Agricultural Organization (Salm 1982), US Agency for International Development (Soekarno 1989), World Wildlife Fund (Djohani et al. 1989), and the Asian Development Bank and the Indonesian Ministry of Forestry (1992).

10. The Togean Islands have been proposed as different kinds of protected areas with different levels of restriction at different moments in time. These have included "Marine Multiple-Use Reserve" (*Kawasan Konservasi Alam Laut*), "Strict Nature Reserve" (*Cagar Alam*), "Marine Sanctuary" (*Suaka Margasatwa Laut*), "National Marine Park" (*Taman National Laut*), and "Natural Recreation Area" (*Taman Wisata Laut*). For the purposes of the ultimate aims of the IFABS organization and the narrative of *Wild Profusion*, I view all of these as forms of "national park."

11. To get a sense of this debate as it has impinged on the issue of biodiversity, see Cronon (1996) and Soulé (1995).

12. For an excellent ethnographic example of a similar project sponsored by World Wide Fund for Nature in Tanzania, see Whalley (2004).

13. In 1680 the crew of the Dutch "hooker" (a kind of ship), the *Brandegans*, made the first map of the Gulf of Tomini. It showed a settlement in the Togean Islands labeled "Negeri Togoya."

14. "Puah" is the Sama language teknonym for a person, female or male, who has children but not yet grandchildren. "Mbo" is the teknonym for one who has become a grandparent.

15. "Sama" is an ethnonym that the people use to refer to themselves. Non-Sama people more often than not use the exonym "Bajau." My research with Togean people was sited in several locations: two different non-Sama Togean villages (Malenge and Katupat); three Sama villages (Pulo Anau, Susunang, and Toani); three Sama hamlets (Dano, Pulo Papan, and Penabali). The work with scientists in the Togean Islands developed primarily at the biological research station on Malenge Island, Camp Uemata.

Part I: Diversity as Milieu

1. As Benedict Anderson argues, it is "political at such a deep level that almost everyone . . . [is] unconscious of the fact" (Anderson 1990:183).

2. Shelly Errington has observed that while the idea of progress has come under suspicion in the North, the idea of cultural evolution has had a new life breathed into it across the global South:

> Governments of third-world (or "developing") countries have tended to embrace the idea of progress with enthusiasm. Originally called "modernization," more recently "development," the idea of ceaseless forward economic and technological movement has been given new life by an alliance composed of authoritarian third-world regimes, transnational corporations, international monetary and development agencies, and consultants from the industrialized state economies. Like early discourses of progress,

these late-twentieth-century avatars invent objects that appropriate and refer to the primitive and the past, although these are more likely to take the form of glorification of national heritage rather than primitive art . . . or cultural theme parks celebrating ethnicity or national history, or decorated and themed airports and hotels. (Errington 1998:5–6)

Chapter 1: Making the Monkey

1. In the Fall/Winter 1994 edition of the *Sulawesi Primate Newsletter*, a contributor from CI wrote, "The presence of the endemic Togean macaque (*Macaca togeanus*) and great significance of the Togians for biogeography and evolutionary biology led to the development of a long-term research conservation and community development program there" (Mackie 1994).

2. This funding was procured through USAID's Biodiversity Conservation Network.

3. Cori Hayden discusses the need to move beyond the sets of instrumental "interests" in science that practitioners of science and technology studies have traditionally followed. Science studiers have followed "interests" to understand how facts become facts. Hayden observes the unfortunate affinity of this method with the "rational actor" model inherent in conservation discourse itself. She writes, "mainstream science studies and neoliberal biodiversity discourse share a fascination, it would seem, with *Homo economicus* and his rational, interest-maximizing behavior" (Hayden 2003:20–23).

4. Due to the compulsory nature of modernity's forms, Chakrabarty has insightfully objected to the term "alternative" (Chakrabarty 2002:xx).

5. See Abraham 2000, 1998, Chakrabarty 2000, Kumar 1995, and Prakash 1999, among others, for work that has explored the relationship between science and Indian and British nationalisms. Paul Rabinow (1999) has explored French nationalism and DNA through his work at the *Centre d'Etude du Polymorphisme Humaine*, while Joan Fujimora (2000) explores nation and genomics in Japan. These are only a few examples of emerging work on science and nation in the field of science and technology studies.

6. Specifically, in Pramoedya's narrative Indies natives are unable to represent themselves in court in the Dutch language they have mastered so painstakingly.

7. Although the army in West Java had demonstrated its capacity for harshness in fighting against the Darul Islam movement, they were also considered to be closer to the peasantry as a legacy of the battalion's role in the struggle for Indonesian independence. For this reason, it has been argued that the West Java battalion did not instigate the same level of killing as other army groups did in 1965 and 1966 (Cribb 1990).

8. This situation is analogous to one Anagnost (2004) describes as the "corporeal politics of quality" in China. In China's urban centers, migrant laborers underwrite China's expanding market-oriented economy. While migrant labor is coded as "low quality" and is devalued as a massive obstacle to modernization, the perceived lack of value in the migrant laborer's body allows a surplus to be extracted that enables capital accumulation. It is the migrant laborer's "derogation," Anagnost argues, that produces the value that will be surplus. The Chinese concept of value, *suzhi*, is very similar to the Indonesian conception of "human

resource quality" that will be added to the national subject through scientific modernization. See also Rutherford (2002) for how similar ideas are applied in West Papua.

9. For a description of the role of Walhi in Indonesian social and environmental change in Suharto-era Indonesia, see Tsing (2004).

10. One controversy that can be found within this science, however, concerns the species concept itself. A major division is between the "biologic" species concept and the "phylogenetic" species concept. The former emphasizes taxonomic relationships while the latter emphasizes evolutionary connections. The controversy has been described this way:

> Scientific theories direct attention to the relationships and interactions among things, or entities, thought to have an ontological status in nature. Theories themselves sometimes impart or predict the expectation of reality to things not known to exist. One only has to recall the predictions of particle physics or astronomy to appreciate this. Likewise, theories of evolution are theories about the descent and modification of entities. Biologists have traditionally called these entities "species," but it is essential to remember that the term species has been used in many different ways. Moreover, it is widely recognized that taxonomic species as such have not always been accepted as the "units of evolution" by many biologists. Indeed, these "units" have been postulated to be genes, gametes, individual organisms, local demes, populations, varieties, subspecies, "biological species," and even higher taxa. If this vast nomenclature signifies anything, perhaps it is the diversity of opinion that exists about the evolutionary process itself, for conceptions or theories about how nature is organized and has developed do influence opinions about the ontological status of the evolving entities. (Cracraft 1992:96)

11. Yet another of these tensions is over the contribution Northern scholars make to Indonesia after they have gathered all their data. One discourse readily available to Indonesian scientists that fuels a great deal of resentment is that Northern scientists take their data and run. Even thought foreign scholars frequently pursue future careers in the EuroAmerican academy, my experience is that many of them spend the rest of their career engaging with Indonesia in a productive and caring way and contributing to a positive representation of their Indonesian experiences.

12. John Pemberton describes his experience of acquiring a research visa in his ethnography of Java:

> When I returned to Indonesia in February 1982 to research what I loosely perceived as the peculiar relationship between culture and politics in Java under the New Order rule of the Soeharto regime, something marvelous happened. The research permit that I had tracked for months through various offices of project sponsorship, security confirmation, and police identification did indeed "emerge" (as Indonesian bureaucrats invariably put it, as if they themselves are pleasantly surprised by such results) on February 19, 1982. But the permit also expired on the same date. (1994:1)

13. Dipesh Chakrabarty expands upon the problem in relation to transnational citational practices. "Third world historians feel a need to refer to works in European history; historians of Europe do not feel the need to reciprocate" (2000:28).

14. Itty Abraham (2000) has written about "site" or location as a "signpost" for science in modern India. He writes that in studies of cosmic rays, Indian scientists used their location near the equator to make up for other disadvantages in funding or equipment. He writes, "Natural features became, willy-nilly, the silent allies of postcolonial science." At the same time, the existence of nonscientific cultural formations, such as temples, villages, and fields were approached as "extrascientific problems" by these same scientists.

15. I thank Jonathan Marks for informing me of the length of evolutionary time separating contemporary humans from macaques.

16. This should not be understood as an "Indonesian" phenomenon, but rather as a "scientific" one. In another context, speaking about a scientific conference that he had attended in France, Paul Rabinow writes, "clearly, the actual event itself was of little or no importance except that it had taken place and could consequently form the basis of a paragraph for a funding proposal submitted to finance the next event" (1996a:4).

Chapter 2: The Social Turn

1. For example, see Harold Conklin (1998). He writes,

> My enthusiasm for things anthropological was sparked on a hot August afternoon in a dusty arena at the 1933 Chicago World's Fair. I was seven years old and entranced by the performance of a large, well-outfitted group of Plains Indian singers, drummers, and dancers. My father, who had brought me along on a week's trip by train from the East Coast, suggested I go down from our seats in the bleachers for a closer look. Delighted, I did not hesitate. A few minutes later a Dakota woman beckoned, and I joined my first Native American friends and, for the next hour, followed their lead and instructions. Their hospitality, rhythms, voices, songs, trappings, colors—even the smell of buckskin and rawhide—I found extremely appealing.

2. Other Indonesian biologists, however, were suspicious that people could live successfully in parks in Indonesia or elsewhere. Those scientists who were the most skeptical of, or uninterested in, Togean people tended to be those with a strict interest in plant and animal biology.

3. It was also necessary for him to earn a living and he sold his collections to Cosimo di Medici III, who kept Rumphius's curiosities in his private *Rariteitkamer* in Tuscany.

4. Not the least of whom was his Ambonese wife Susanna, who helped him to discover plants and after whom he named a rare orchid *flos Susanna* (Beekman 1999:lxvii).

5. In the 1870s, Meyer shipped several samples of five species of birds he had captured in the Togean Islands to Viscount Arthur Walden, otherwise known as Arthur Hay, the Ninth Marquess of Tweeddale, President of the Zoological Society of London from 1868 to 1878. In Walden's accounts (1871, 1872) narrative natures morphed into pure morphology. In Walden's entry on the *Loriculus quadricolor,* he described this new species of lorikeet from the Togean Islands in the following manner:

Loriculus quadricolor, n.sp.

Adult male. Bright green; crown and edge of shoulder scarlet; rump, upper tail-coverts, chin and throat deep blood-red; interscapulars back bright golden; quills black, half the inner web of each quill verditer blue; entire undersurface of rectrices verditer blue; bill black, feet yellow.

Male, immature. Faint indications of a few scarlet feathers on the forehead; a small red spot on the throat; edge of the shoulder scarlet mixed with yellow; upper tail-coverts and rump and remainder of plumage as in the adult. This stage closely resembles the adult plumage of *L. Wallacei*, G. R. Grey.

Female. The only example sent and thus marked by Dr. Meyer is not quite adult. The head is entirely green, the chin and throat scarlet, the shoulder-edge yellow, interscapulars golden, remainder of plumage as in adult male.

	Rostr. a nar.	*Alae.*	*Caudae.*	*Tarsi.*
Adult m.	0.37	3.69	1.88	0.37
Adult f.	0.31	3.50	1.63	0.37

This species is intermediate between *L. stigmatus* (Muller and Schlegel) and *L. Wallacei*, G. R. Grey. In dimensions the three are about equal. From *L. Wallacei* it differs by having a scarlet cap, by the golden of the back reaching to the nape, by the darker red of the uropygium and upper tail-coverts, and by the sexes differing; from *L. stigmata* by the golden back, by the chin and throat-spot being much smaller, and the red of the uropygium not being quite so dark. All the examples sent are from the Togian Islands. (Walden 1872)

6. The Resident of Manado in 1878, S.C.J.W. Van Musschenbroek (1880), describes Meyer's travels in Minahassa and Tomini, and Meyer's work with the parrots with great praise. He writes that everyone knew of the importance of these parrots, but they had been treated in the wrong way, unsystematically. Meyer's good result was only possible due to his large sample size. All the museums of Europe could not have given him such samples.

7. For example, to the question, "How large should a park be?" John Terborgh answers, "As a rough guide, the area should be the size needed to sustain genetically viable populations of top predators" (Terborgh 1999:62).

8. The earliest survey to propose a spatial conservation strategy for the Togean Islands was performed by the Asian Development Bank in 1982 (1982 Salm et al.). This survey identified the Togean Islands for a variety of "interests and values," including reefs and endangered species, and the presence of "interesting Sea Gypsy (Bajo) stilted villages." Three maps in the report proposed territorial boundaries demarcating a reserve boundary around the islands, and a zonation system within the proposed protected area. This zonation system would have closed off the most frequented fishing grounds in the archipelago.

9. From an environmental justice perspective, there are many potential political dilemmas in projects of counter-mapping. Hodgson and Schroeder describe some of them:

> Conflicts inherent in conservation efforts involve territorialization, privatization, integration and indigenization; problems associated with the theory and practice of "community-level" political engagement; the limited scope of two-dimensional mapping tools; the need to combine mapping efforts with broader legal and political strategies; and critical questions involving the agency of "external" actors such as conservation and development donors, the state, and private business interests. (Hodgson and Schroeder 2002:82)

10. In his report on nutmeg, for example, Rumphius wrote of the number of slaves that died under the forced labor of the nutmeg harvest, and he revealed how native planters were required to burn their surplus crop.

11. The only real moment of danger came when she related the story to her colleagues or to me. If we were to insist upon these divides, either between Laksmi and the farmer, or between ourselves and Laksmi, then Laksmi, perhaps, would herself have been forced to take a more realist position.

Chapter 3: Extraterrrestrial Others

1. Beginning in the eleventh century, Samalan speakers began to migrate out of the southern Philippines in search of marine products, mainly sea cucumber and sharks fin, which they sold to land-based rulers for trade with China. They worked their way down the Sulu archipelago from Mindanao and through Borneo, reaching the Makassar Strait and spreading out into eastern Indonesia before the colonial period. See James Fox (1993).

2. Van der Wal described his first encounter with Togean people in Kilat Bay:

> We learned from these people it was *Negri Togoya*, located on the south side of the Island. It was ruled by a Queen. But from this side, one could only get there with a small canoe, like theirs. And they said if we wanted to trade, they had turtle shell and mouse foods. We would have to sail to the south side where they would then meet us with large vessels. . . . Toward evening some of the inhabitants came to ask us what we wanted and why we were here. We answered we came for trade if they had any turtle shell. They said they didn't have any now but if we wanted to stay the King would speak with us. In the morning, the King came with the *Gugugu* and the Sea Captain and 60 armed men. They came down the mountain bringing us coconuts, betelnut, and chickens, gifts which I accepted. The Sea Captain who spoke for the King asked what kind of people we were and from where. (van der Wal 1680)

3. The approximate date and location of early Sama settlements are confirmed by Rili Djohani (1993) who writes that Sama people told her they came to the Togeans in the 1860s from the Salabanka Islands and first settled at Pulo Anau.

4. Two streams of cognitive anthropology have entered into these debates from opposite ends of the spectrum. One stream, sometimes referred to as "ethnological" ethnoscience, has used folk categories to speak of universals in nature and language. This work examines the similarities in categorizations across cultures. Brent Berlin (1992), for example, bases his work on what he sees as similarities between systematic biology and folk biology, which both reflect, in his mind, the

"natural system" of nature. He finds it conclusive that if a bunch of Amazonian bird skins are dumped on a table, students, ornithologists, and native peoples will all group the birds in the same way. The second stream, the "ethnographic" ethnoscience, uses folk taxonomies to argue for cultural specificity. If one dumped a basket of the 1,500 or so plant species found by Harold Conklin (1957) in Hanunóo onto the same table, it would be impossible to uncover the same universal natural system and categorizations would depend on who you are. From this perspective, nature is known through standpoints.

5. Many have written of the storied and experiential nature of Southeast Asian landscapes. Here I am in dialogue with Doolittle (2005), Dove (1986), George (1996), Harwell (2000), Li (1999), Peluso (1996), Sturgeon (2005), Rosaldo (1980), Roseman (2003), Tsing (1993, 2004), and Zerner (2003). While not specifically about Southeast Asia, I have also found Kathleen Stewart's (1996) description of West Virginia especially inspiring here. James Fox (1977) should be consulted for the complete story of the sago palm in eastern Indonesia.

6. Henk Maier (1993) asks the provocative question, "How did the inhabitants of the Indies know that they were speaking more than one language? Were they aware of it, at all, and even more: were they aware of differences between the two languages, or three languages they were using, and of their concurrent affectivity?"

7. For the argument against "experience" as the foundation of identity see Scott (1992). For a recuperation of the idea of experience see Moya and Hames-Garcia (2000).

Chapter 4: On the (Bio)logics of Species and Bodies

1. While "siang" is Indonesian for midday, Togean Sama people use the Indonesian word "siang" to refer to the time immediately before daylight. Thus, "hampir siang" means almost daylight.

2. Through her work on atherosclerosis, Annemarie Mol (2002:32–36) tackles the thorny problem of subject-object relations by introducing the concept of "enactment." Describing objects as "enacted," she argues, allows us to see that "things" don't have a life of their own independent of the practices that enliven them. As such, objects require techniques to make them tangible, visible, audible, and ultimately knowable. The implication of this is that objects, once produced, do not then take on a stable existence now independent of practices. New practices beget new objects, and objects continually require techniques for their instantiation.

3. Despite the contention of some science studies scholars that we must abandon the "imagination" in favor of empirical practices, I prefer to track both simultaneously. Robert Romanyshyn (1989) makes an argument for the role of "technology as cultural-historical dream" informing our sense of self. He writes: "We may, perhaps, state the point this way: technology is a crisis of the imagination. As such, technology is both a danger and an opportunity with respect to the imagination. It is a danger in so far as it can be the death of imagination through its literalization. It is an opportunity in so far as it can be an awakening to how the events of the world have an imaginal depth, and how the life of the imagination inscribes itself within the events of the world."

4. James Scott (1998) has developed an important argument concerning "how certain schemes to improve the human conditions have failed." Scott argues that "certain forms of knowledge and control require a narrowing of vision." What he terms "state simplifications" define the narrowing of vision and purpose used to achieve political order and managerial control, and he notes the similarity between these state endeavors and the goals of large corporations. His analysis is not antithetical to an analysis of care. Through an analytic of care, however, it would not be the "forms of knowledge" that demand a narrowing of vision, but the ends the knowledge serves. A narrowing of vision through systems of classification applied to regulating the safety standards of public transportation, for example, would have the capacity to "improve the human condition."

Chapter 5: Fishing with Cyanide

1. Following Appadurai (1995 [1986]), I have divided commodity knowledge into categories of production and consumption. Using the example of bird nests collected by people who live in forests in Borneo, he notes the gap producers have in their knowledge of market destinations. He claims this allows traders high profits and depends on the subordinance of the producing region relative to the consuming one. Kopytoff (1995) argues that the knowledge gap also prevents producing and consuming cultures from having their understanding of commodities challenged.

2. For a further elaboration of environmental laws in Indonesia see Koesnadi Hardjasoemantri (1991), and Carol Warren and Kylie Elston (1994). Warren and Elston argue that Indonesian legal regulation is "studded with 'special dispensations' and selective application according to 'vulnerability and political value' " (1994:8).

3. This decree falls under the legislative aegis of the Director General of Fisheries within the Ministry of Agriculture.

4. Article 6 of the Decree of the Director General of Fisheries No SK.330/DJ.8259/95 (GOI 1995b) states "the collecting company shall provide means for cultivation at the stated collecting location and possess a staff experienced in fish cultivation." The Napoleon wrasse cannot be spawned successfully in captivity, however.

5. With the right permit, compressors are not illegal, even though they still would be most productively employed, from an owner's perspective, in catching live fish with cyanide.

Chapter 6: The Sleep of Reason

1. To understand the non-prediscursive nature of nature informing the difference of Indonesians' biodiversity science, it is useful to reflect on Judith Butler's disruption of the "matter" of bodies, especially woman's bodies.

> The body posited as a sign, is always *posited*, or *signified*, as *prior*. This signification produces as an *effect* of its own procedure the very body that it nevertheless and simultaneously claims to discover as that which *precedes* its own action. If the body signified

> as prior to signification is an effect of signification, then the mimetic or representational status of language, which claims that signs follow bodies as their necessary mirrors, is not mimetic at all. On the contrary it is productive, constitutive, one might even argue *performative*, inasmuch as this signifying act delimits and contours the body that it then claims to find prior to any and all signification. (Butler 1993:30)

2. This sentiment is now expressed quite widely in conservation circles. For example, it was a major theme of a talk on ecoregions by Gordon Orians, emeritus board member of World Wildlife Fund, given at University of Washington in 2001. It was also an argument made by several biologists from CI and TNC at the conference "Conservation for/by Whom? Social Controversies & Cultural Contestations Regarding National Parks and Reserves in the 'Malay Archipelago' " at Singapore National University, May 2005.

3. The CI report frames the need this way: "Effective enforcement in hotspots will require much broader action on multiple fronts, including augmenting inter-agency cooperation; increasing enforcement agency budgets; building technical capacity of detection agents, prosecutors, and judges; implementing enforcement performance monitoring systems; and strengthening natural resource laws and internal policies of enforcement agencies" (CI 2004:2).

4. Nicholas Rose describes these politics as "minor engagements": "[O]ne would examine the ways in which creativity arises out of the situation of human beings engaged in particular relations of force and meaning, and what is made out of the possibilities of that location. These minor engagements do not have the arrogance of programmatic politics—perhaps even refuse their designation as politics at all. They are cautious, modest, pragmatic, experimental, stuttering, tentative" (Rose 1999:279–80).

5. Some might want to argue this is *their* nightmare, too, of the world of species extinction. The sea cucumbers, the giant clams, the macaques all have gone missing in this apocalyptic image. Boi is left with nothing because an over-populous "humanity" has exceeded its carrying capacity. This is the future of humanity once we have used up all its nature and extinction is the order of the day. I would counter that since the list of species biologists want to protect overlaps with nearly every plant or animal important to Sama people, as it did in the Togean Islands, then the interests of transnational biodiversity conservation have never included a true vision of Sama well-being. A program designed with Sama futures in mind might try and boost the numbers of sea cucumber or giant clams, or protect them from rapacious industrial resource extraction. The apocalyptic vision of transnational biodiversity conservation is one where a scientist has nothing left to study, or an ecotourist has nowhere left to travel, not the vision of Boi scavenging on the Dumaguete beach, which is naturalized as external to the task of saving biodiversity.

6. Pheng Cheah makes this argument by first disentangling instrumental from critical reason in the work of Pramoedya Ananta Toer on Indonesia's colonial awakening. Toer believes, "instrumentality is merely an accident that befalls reason, a removable stain external to reason in the final instance, because it is a consequence of how reason is used" (Cheah 2003:302). I believe Cheah envisions these two forms of reason as significantly entangled.

References

Abraham, Itty. 2000. "Postcolonial Science, Big Science, and Landscape." In *Doing Science + Culture*. Roddey Reid and Sharon Traweek, eds. Pp. 49–70. London: Routledge.

———. 1998. *The Making of the Indian Atomic Bomb: Science, Secrecy, and the Postcolonial State*. London: Zed Books.

Adriani, N. 1900. "De Talen der Togian-eilenden," *Tijdschrift voor Indische Taal-, Land-, en Volkenkunde*, 42:46, 490, 539–66.

Agrawal, Arun. 1995. "Dismantling the Divide Between Indigenous and Western Knowledges." *Development and Change*. 26(3): 413–39.

Anagnost, Ann. 2004. "The Corporeal Politics of Quality (*Suzhi*)." *Public Culture* 16(2):189–208.

Anderson, Benedict. 1998. *The Spectre of Comparisons: Nationalism, Southeast Asia, and the World*. London: Verso.

———. 1990. "Cartoons and Monuments: The Evolution of Political Communication Under the New Order." In *Language and Power: Exploring Political Cultures in Indonesia*. Ithaca: Cornell University Press.

Appadurai, Arjun. 1995. (1986) "Introduction: Commodities and the Politics of Value." In *The Social Life of Thing: Commodities in Cultural Perspective*. Arjun Appadurai ed. Cambridge: Cambridge University Press.

Aragon, Lorraine. 2000. *Fields of the Lord: Animism, Christian Missionaries, and State Development in Indonesia*. Honolulu: University of Hawaii Press.

Asian Development Bank and Indonesian Ministry of Forestry. 1992. Management and Conservation of Tropical Forest Ecosystems and Biodiversity. Project T.A. 1430-INO. Deutch Forst Consult, Germany, and PT Hasfarm Dian Konsultan, Jakarta.

Atkinson, Jane. 1996. "Quizzing the Sphinx: Reflections on Mortality in Central Sulawesi." In *Fantasizing the Feminine in Indonesia*. Laurie Sears, ed. Durham: Duke University Press.

Banuri, T. and F. Apffel-Marglin, eds. 1993. *Who Will Save the Forests?: Knowledge, Power, and Environmental Destruction*. London: Zed Books.

Barkan, Elazar, and Ronald Bush. 1995. *Prehistories of the Future: The Primitivist Project and the Culture of Modernism*. Stanford: Stanford University Press.

Basso, Keith. 1996. *Wisdom Sits in Places: Landscape and Language Among the Western Apache*. Albuquerque: University of New Mexico Press.

Beekman, E. M. 1999. "Introduction: Rumphius' Life and Work." In *The Ambonese Curiosity Cabinet*. Georgius Everhardus Rumphius. New Haven: Yale University Press.

———. 1981. *The Poison Tree: Selected Writings of Rumphius on the Natural History of the Indies*. Amherst: University of Massachusetts Press.

Berlin, Brent. 1992. *Ethnobiological Classification*. Princeton: Princeton University Press.

Bhabha, Homi K. 1994. *The Location of Culture*. London: Routledge.

Bourdieu, Pierre. 1998. *Practical Reason*. Stanford: Stanford University Press.

Bourdieu, Pierre. 1984. *Distinction: A Social Critique of the Judgment of Taste.* Cambridge: Harvard University Press.

Burchell, Graham, Colin Gordon, and Peter Miller. 1991. *The Foucault Effect: Studies in Governmentality.* Chicago: University of Chicago Press.

Bush, Ronald, and Elazar Barakan. 1995. *Prehistories of the Future: The Primitivist Project and the Culture of Modernism.* Stanford: Stanford University Press.

Butler, Judith. 1993. *Bodies that Matter: On the Discursive Limits of "Sex."* London: Routledge.

Bynum, Eva L. 1994. "Conservation Status of Sulawesi Macaques." Sulawesi Primate Newsletter. 2(2):7.

Chakrabarty, Dipesh. 2002. *Habitations of Modernity: Essays in the Wake of Subaltern Studies.* Chicago: University of Chicago Press.

———. 2000. *Provincializing Europe: Postcolonial Thought and Historical Difference.* Princeton: Princeton University Press.

Chatterjee, Partha. 1993. *The Nation and its Fragments: Colonial and Postcolonial Histories.* Princeton: Princeton University Press.

Cheah, Pheng. 2003. *Spectral Nationality: Passages of Freedom from Kant to Postcolonial Literatures of Liberation.* New York: Columbia University Press.

Colfer, Carol. 2002. "Ten Propositions to Explain Kalimantan's Fires." In *Which Way Forward?: People, Forests, and Policymaking in Indonesia.* Washington: Resources For the Future Press.

Conklin, Harold. 1998. "Language, Culture, and Environment: My Early Years." *Annual Review of Anthropology* 27:xiii–xxx.

———. 1957. "Hanunóo Agriculture: A Report on an Integral System of Shifting Cultivation in the Philippines." FAO Forestry Development Paper No. 12. Rome: FAO.

Conservation International. 2004. New Study Shows That When it Comes to the Environment, Crime Still Pays. Hotspots E-News. *http://www.maildogmanager.com/page.html?p=000001XDDtjF4u9sQYkwYguzLx4+kB1mFXrQ* (accessed 8/8/05).

Cracraft, Joel. 1992. "Species Concepts and Speciation Analysis." In *The Units of Evolution: Essays on the Nature of Species.* Marc Ereshefky, ed. Pp. 28–59. Cambridge: MIT Press.

Cribb, Robert. 1990. *The Indonesian Killings of 1965–66: Studies from Java and Bali.* Clayton, Victoria: Centre of Southeast Asian Studies, Monash University.

Cronon, William, ed. 1996. *Uncommon Ground: Rethinking the Human Place in Nature.* New York: W.W. Norton.

Darwin, Charles. 1874. *The Structure and Distribution of Coral Reefs.* London: Smith, Elder.

Daws, Gavan, and Marty Fujita. 1999. *Archipelago: The Islands of Indonesia.* Berkeley: University of California Press.

Djohani, Rili Hawari. 1989. *Marine Conservation Development of Indonesia: Coral Reef Policy.* Jakarta: WWF.

Djohani, Rili, Ramli Malik and Muslimin. 1993. "Marine Parks and the Socioeconomic Implications for the Bajau People, Particularly from the Togian Islands." Prepared for the International Seminar on Bajau Communities, Jakarta.

Doolittle, Amity. 2005. *Property and Politics in Sabah Malaysia: Native Struggles Over Land Rights*. Seattle: University of Washington Press.

Dove, Michael. 1986. "The Practical Reason for Weeds: Peasant vs. State Views of Imperata and Chromoleana." *Human Ecology* 14(2):163–90.

Duncan, Christopher, ed. 2004. *Civilizing the Margins: Southeast Asian Government Policies for the Development of Minorities*. Ithaca: Cornell University Press.

Ellen, Roy, Peter Parkes, and Alan Bicker. 2000. *Indigenous Environmental Knowledge and its Transformations*. Amsterdam: Harwood Academic Publishers.

Erdmann, Mark. 2002. "Perspective: The WAR on Destructive Fishing Practices." SPC Live Reef Fish Information Bulletin #10. Noumea: Secretariat of the Pacific Community.

———. 2001. "Who's minding the reef? Corruption and Enforcement in Indonesia." SPC Live Reef Fish Information Bulletin #8. Noumea: Secretariat of the Pacific Community.

Errington, Joseph. 1998. *Shifting Languages: Interaction and Identity in Javanese Indonesia*. Cambridge: Cambridge University Press.

Errington, Shelly. 1998. *The Death of Authentic Primitive Art and Other Tales of Progress*. Berkeley: University of California Press.

Fabian, Johannes. 2000. *Out of Our Minds: Reason and Madness in the Exploration of Africa*. Berkeley: University of California Press.

Faisal, Agus. 2001. "Togean People Kick Out Imperialists." *World Rainforest Movement Bulletin*. No. 52., November 2001.

Fanon, Frantz. 1991(1952). *Black Skin, White Masks*. New York: Grove Books.

Feld, Steven. 1982. *Sound and Sentiment: Birds, Weeping, Poetics, and Song in Kaluli Expression*. Philadelphia: University of Pennsylvania Press.

Firman. 1994. "Studies on Population and Diet of the Togean Macaque (*Macaca Togeanus*)." Sulawesi Primate Newsletter 2(2):7.

Fooden, Jack. 1969. *Taxonomy and Evolution of the Monkeys of Celebes*. Basel: S. Karger.

Foster, Nancy. 1998. Keynote Address: "Marine Protected Areas in the New Millennium." Proceedings of the International Tropical Marine Ecosystems Management Symposium, November 22–26. Port Townsend: Great Barrier Reef Marine Park Authority.

Foucault, Michel. 1991. "Governmentality." In *The Foucault Effect: Studies in Governmentality*. Graham Burchell, Colin Gordon, and Peter Miller, eds. Chicago: University of Chicago Press.

———. 1975. *Discipline and Punish: The Birth of the Prison*. Alan Sheridan, trans. New York: Vintage Books.

———. 1972. *The Order of Things: An Archaeology of the Human Sciences*. New York: Vintage Books.

Fox, James. 1993. "Bajau Voyages to the Timor Area, the Ashmore Reef, and Australia." Indonesian Institute of Sciences, Jakarta, November 22–25.

———. 1977. *Harvest of the Palm: Ecological Change in Eastern Indonesia*. Cambridge: Harvard University Press.

Franklin, Sarah, and Susan McKinnon. 2001. *Relative Values: Reconfiguring Kinship Studies*. Durham: Duke University Press.

Freud, Sigmund. 1989 [1933]. *New Introductory Lectures on Psycho-Analysis*. New York: Norton.

Froehlich, Jeffrey W., Jatna Supriatna, V. Hart, S. Akbar, and R. Babo. 1998. "The Balan of Balantak: A Possible New Species of Macaque in Central Sulawesi." *Tropical Biodiversity* 5(3):167–94.

Froehlich, Jeffrey W., Jatna Supriatna. 1996. "Secondary Intergradation Between *Macaca maurus* and *M. tonkeana* in South Sulawesi, and the species status of *M. togeanus*." In *Evolution and Ecology of Macaque Societies*. John Fa and Donald Lindburg, eds. Pp. 43–70. Cambridge: Cambridge University Press.

Fujimora, Joan. 2000. "Transnational Genomics: Transgressing the Boundary Between the "Modern/West" and the "Premodern/East." In *Doing Science + Culture*. Roddey Reid and Sharon Traweek, eds. Pp. 71–92. London: Routledge.

George, Kenneth. 1996. *Showing Signs of Violence: The Cultural Politics of a Twentieth-Century Headhunting Ritual*. Berkeley: University of California Press.

Gilligan, Carol. 1982. *In a Different Voice: Psychological Theory and Women's Development*. Cambridge: Cambridge University Press.

Government of Indonesia. 2004a. Decree of the Minister of Forestry Number: SK.418/Menhut-II/2004.

———. 2004b. Law No. 32/2004 Concerning Regional Governance.

———. 1999a. Law No. 22/1999 Concerning Regional Governance.

———. 1999b. Law No. 31/1999 Concerning Fisheries.

———. 1995a. Decree of the Agrarian Minister of the Republic of Indonesia. No. 375/Kpts/JK.250/5/95 Regarding the Ban on Napoleon Wrasse Fish Haul.

———. 1995b. Decree of the Director General of Fisheries. No. SK330/DJ.8259/95 Regarding Size, Location, and Manners of Hauling Napoleon Wrasse.

———. 1995c. Decree of the Trade Minister of the Republic of Indonesia. No. 94/Kp/V/95 Regarding the Ban on Export of Napoleon Wrasse Fish.

———. 1992. Law 24/1992: Spatial Planning.

———. 1990. Law No. 5 1990 About the Conservation of Natural Resources and Ecosystems.

Grove, Richard. 1995. *Green Imperialism: Colonial Expansion, Tropical Island Edens, and the Origins of Environmentalism*. Cambridge: Cambridge University Press.

Gupta, Akhil. 1998. *Postcolonial Developments: Agriculture in the Making of Modern India*. Durham: Duke University Press.

Gupta, Akhil, and James Fergusson. 1997. *Culture, Power, Place: Explorations in Critical Anthropology*. Durham: Duke University Press.

Hankivsky, Olena. 2004. *Social Policy and the Ethic of Care*. Vancouver: University of British Columbia Press.

Haraway, Donna. 1991. *Simians, Cyborgs, and Women: The Reinvention of Nature*. New York: Routledge.

———. 1989. *Primate Visions: Gender, Race, and Nature in the World of Modern Science*. New York: Routledge.

Harding, Sandra. 1998. *Is Science Multicultural: Postcolonialisms, Feminisms, and Epistemologies*. Bloomington: University of Indiana Press.

Hardjasoemantri, Koesnadi. 1991. *Hukum Perlindungan Lingkungan: Konservasi Sumber Daya Alam Hayati dan Ekosystemnya* (*Law for Environmental Preservation: Conservation of Biological Resources and Their Ecosystems*). Yogyakarta: Gajah Mada University Press.

Hartsock, Nancy. 1999. *The Feminist Standpoint Revisited and Other Essays*. Boulder: Westview Press.

Harwell, Emily. 2000. "Remote Sensibilities: Discourses of Technology and the Making of Indonesias Natural Disaster." *Development and Change*. 31(1):307–340.

Hayden, Cori. 2003. *When Nature Goes Public: The Making and Unmaking of Bioprospecting in Mexico*. Princeton: Princeton University Press.

Head, Jonathan. 1998. Profile: President B. J. Habibie. BBC Online News. May 21, 1998.

Hefner, Robert. 1990. The Political Economy of Mountain Java: An Interpretive History. Berkeley: University of California Press.

Hodgson, Dorothy, and Richard A. Schroeder. 2002. "Dilemmas of Counter-Mapping Community Resources in Tanzania." *Development and Change*. 33(2002):79–100.

van Hoevell, G.W.W.C. Baron. 1893. *Bijschrift bij de Kaart der Tomini-Bocht* (*Explanation of the Tomini Gulf Map*). Leiden: E.J.Brill.

Horkheimer, Max, and Theodor Adorno. 2001 [1944]. *Dialectic of Enlightenment*. New York: Continuum.

Hussey, Edward B. Jr. 1855. *Log of the Ship Peruvian,* unpublished ship's log (Log #53). Mystic Seaport, Connecticut.

Hutabarat, C., Pramono, A.H., Yuliati, S. 1996. "Preliminary Study on Marine Resources Used in Togian Islands, Sulawesi (Indonesia): With Special Reference to Bajau People." *Ethnobiology in Human Welfare*. New Delhi: Deep Publications. Pp. 447–50.

Indonesian Foundation for the Advancement of Biological Sciences. 1997. Concept Paper: "Participatory Spatial Planning in Togean Islands." Jakarta: IFABS.

Indrawan, Didi M. 1997. "Notes on Togean Conservation." Unpublished notes.

International Crisis Group. 2001. "Indonesia: Natural Resources and Law Enforcement." ICG Asia Report #29, Jakarta/Brussels.

Johannes, Robert E. and Michael Riepen. 1995. "Environmental, Economic, and Social Implications of the Live Fish Trade in Asia and the Western Pacific." Consultant's Report to The Nature Conservancy and The South Pacific Forum Fisheries Agency.

Kant, Immanuel. 2001.[1781]. *Basic Writings of Kant*. Allen Wood, ed. New York: The Modern Library.

Kantor Statistik Kebupaten Poso. 1995a. *Kecamatan Una Una Dalam Angka*. Poso: Kantor Statistik Kebupaten Poso.

———. 1995b. *Kecamatan Walea Kepulauan Dalam Angka*. Poso: Kantor Statistik Kebupaten Poso.

Keane, Webb. 1997. *Signs of Recognition: Powers and Hazards of Representation in an Indonesian Society*. Berkeley: University of California Press.

Keck, Margaret, and Kathryn Sikkink. 1998. *Activists Across Boarders*. Ithaca: Cornell University Press.
Kopytoff, Igor. 1995 (1986). "The Cultural Biography of Things: Commoditization as Process." In *The Social Life of Thing: Commodities in Cultural Perspective*. Arjun Appadurai ed. Cambridge: Cambridge University Press.
Kumar, Deepak. 1995. *Science and the Raj, 1857–1905*. Delhi: Oxford University Press.
Lakoff, George. 2004. *Don't Think of an Elephant: Know Your Values and Frame the Debate*. White River Junction: Chelsea Green Press.
———. 2002. *Moral Politics: How Liberals and Conservatives Think*. Chicago: University of Chicago Press.
———. 1995. "Metaphor, Morality, and Politics: Or, Why Conservatives Have Left Liberals in the Dust." *Social Research* 62:2 (summer 1995).
Langford, Jean. 2002. *Fluent Bodies: Ayurvedic Remedies for Postcolonial Imbalance*. Durham: Duke University Press.
Latour, Bruno. 1999. *Pandora's Hope: Essays on the Reality of Science Studies*. Cambridge: Harvard University Press.
———. 1987. *Science in Action: How to Follow Scientists and Engineers through Society*. Cambridge: Harvard University Press.
Latour, Bruno, and Stephen Woolgar. 1986. *Laboratory Life: The Social Construction of Facts*. Princeton: Princeton University Press.
Lev, Daniel. 1998. "Religion and Politics in Indonesia and Malaysia," Presentation to the Council on Southeast Asian Studies, Berkeley, California. February 13, 1998.
Lewis, Martin, and Karen Wiggen. 1997. *The Myth of Continents: A Critique of Metageography*. Berkeley: University of California Press.
Li, Tania. 1999. "Marginality, Power, and Production: Analyzing Upland Transformations." In *Transforming the Indonesian Uplands*. Tania Li, ed. Amsterdam: Harwood Academic Publishers.
Lowe, Celia. 2004. "Making the Monkey: How the Togean Macaque went from 'New Form' to 'Endemic Species' in Indonesians' Conservation Biology," *Cultural Anthropology* 19(4):491–516.
———. 2003. "Sustainability and the Question of 'Enforcement' in Integrated Coastal Management: The Case of Nain Island, Bunaken National Park, *Jurnal Pesisir dan Kelautan* (Indonesian Journal of Ocean and Coastal Management) special theme issue 1:49–63.
———. 2003. "The Magic of Place: Sama at Sea, on Land, in Sulawesi, Indonesia," *Bijdragen tot de Taal, Land, en Volkenkunde* 159(1):109–33.
———. 2000. "Global Markets, Local Injustice in Southeast Asian Seas: The Live Fish Trade and Local Fishers in the Togean Islands of Sulawesi, Indonesia." In *People, Plants, and Justice: The Politics of Nature Conservation*. Charles Zerner, ed. New York: Columbia University Press.
Mackie, Cynthia. 1994. "Ecosystem Conservation in the Togean Islands." *Sulawesi Primate Newsletter* 2(2):7.
MacKinnon, J. 1982. "Management and Conservation of Tropical Forest Ecosystems." Asian Development Bank Report T.A. No. 1430–1NO.

Maier, Henk. 1993. "From Heteroglossia to Polyglossia: The Creation of Malay and Dutch in the Indies." *Indonesia* 5:37–66.

Marcus, George. 1995. "Ethnography in/of the World System: The Emergence of Multi-Sited Ethnography." *Annual Review of Anthropology* 24:95–117.

Meyer, Adolf Bernard. 1879. "Field-notes on the Birds of Celebes." *Ibis* (4)3:43–70, 125–47.

Mol, Annemarie. 2002. *The Body Multiple: Ontology in Medical Practice.* Durham: Duke University Press.

Moya, Paula and Michael Hames-Garcia. 2000. *Reclaiming Identity: Realist Theory and the Predicament of Postmodernism.* Berkeley: University of California Press.

Mrazek, Rudolf. 2002. *Engineers of Happy Land: Technology and Nationalism in a Colony.* Princeton: Princeton University Press.

Nehamas, Alexander. 2001. "The Sleep of Reason Produces Monsters." *Representations* 74:37–54.

Nietzsche, Friedrich. 1998[1887]. *On the Genealogy of Morality.* Indianapolis: Hacket.

Nimmo, Harry. 1994. *Songs of Salanda and Other Stories of Sulu.* Manila: Ateneo de Manila Press.

Ong, Aihwa. 1987. *Sprirts of Resistance and Capitalist Discipline: Factory Women in Malaysia.* New York: State University of New York Press.

Peet, Richard, and Michael Watts, eds. 1996. *Liberation Ecology: Environment, Development, Social Movements.* London: Routledge.

Peluso, Nancy. 1996. "Fruit Trees and Family Trees in an Anthropogenic Forest: Ethics of Access, Property Zones, and Environmental Change in Indonesia," *Comparative Studies in Society and History* 38: 510–48.

———. 1995. "Whose Woods Are These?: Counter-mapping Forest Territories in Kalimantan, Indonesia." *Antipode* 27(4):383–406.

———. 1992. *Rich Forests, Poor People: Resource Control and Resistance in Java.* Berkeley: University of California Press.

Pemberton, John. 1994. *On the Subject of "Java."* Ithaca: Cornell University Press.

Pickering, Andrew. 1995. *The Mangle of Practice: Time, Agency, and Science.* Chicago: University of Chicago Press.

Pigg, Stacy Leigh. 1996. "The Credible and the Credulous: The Question of 'Villager's Beliefs' in Nepal." *Cultural Anthropology* 11(2):160–201.

Prakash, Gyan. 1999. *Another Reason: Science and the Imagination of Modern India.* Princeton: Princeton University Press.

Pratasik, Silvester Benny. 1983. *Pengaruh NaCN Terhadap Kehidupan dan Kualitas Daging Ikan Mas dan Ikan Mujair* (The Influence of NaCN on the Life and Tissue Quality of Carp Fish and *Mujair* Fish). Manado, Indonesia: Universitas Sam Ratulangi, Faculty of Fisheries.

Pratt, Mary Louise. 1992. *Imperial Eyes: Travel Writing and Transculturation.* London: Routledge.

Rabinow, Paul. 2003. *Anthropos Today: Reflections on Modern Equipment.* Princeton: Princeton University Press.

———. 1999. *French DNA: Trouble in Purgatory.* Chicago: Chicago University Press.

———. 1996a. *Making PCR: A Story of Biotechnology.* Chicago: University of Chicago Press.

———. 1996b. *Essays on the Anthropology of Reason.* Princeton: Princeton University Press.

Revius. 1852. *Report of Trade In and Between the Islands on the East Coast of Celebes, Bangaaij, and Soella Archipelago.* Republic of Indonesia: Arsip Nasional.

Romanyshyn, Robert. 1989. *Technology as Symptom and Dream.* London: Routledge.

Rosaldo, Renato. 1980. *Ilongot Headhunting: 1883–1974.* Stanford: Stanford University Press.

Rose, Nicholas. 1999. *Powers of Freedom: Reframing Political Thought.* Cambridge: Cambridge University Press.

Roseman, Marina. 2003. "Singers of the Landscape: Song, History, and Property Rights in the Malaysian Rainforest." In *Culture and the Question of Rights: Forests, and Coasts, and Seas in Southeast Asia.* Charles Zemer, ed. Durham: Duke University Press.

von Rosenburg, Carl Benjamin Hermann. 1865. *Reistogten in de Afdeeling Gorontalo, Gedaan op Last der Nederlandsch Indische Regering.* Amsterdam: F. Muller.

Rumphius, Georgius Everhardus. 1999 (1705). *The Ambonese Curiosity Cabinet.* E. M. Beekman, trans. New Haven: Yale University Press.

———. 1981 (1741). Selections from "The Ambonese Herbal." In *The Poison Tree: Selected Writings of Rumphius on the Natural History of the Indies.* Amherst: University of Massachusetts Press.

Rutherford, Danilyn. 2002. *Raiding the Land of the Foreigners: The Limits of the Nation on an Indonesian Frontier.* Princeton: Princeton University Press.

Salm, Rodney, Yulheri Abas, and Rolex Lameanda. 1982. *Marine Conservation Potential: Togean Islands, Central Sulawesi.* Bogor: Food and Agriculture Organization of the United Nations.

Sather, Clifford. 1997. *The Bajau Laut: Adaptation, History, and Fate in a Maritime Fishing Society of South-Eastern Sabah.* Oxford: Oxford University Press.

Sauer, Carl O. 1981. "The Agency of Man on the Earth." In *Selected Essays: 1963–1975.* Carl Sauer, ed. Berkeley: Turtle Island Foundation.

Scott, James. 1998. *Seeing Like a State: How Certain Schemes to Improve the Human Condition Have Failed.* New Haven and London: Yale University Press.

Scott, Joan. 1992. "Experience." In *Feminists Theorize the Political.* Judith Butler and Joan Scott, eds. London: Routledge.

Sears, Laurie J. 1996. "Fragile Identities: Deconstructing Women and Indonesia." In *Fantasizing the Feminine in Indonesia.* Laurie J. Sears, ed. Durham: Duke University Press.

Sejati Foundation. 1994. *Bajau.* Jakarta: Sejati Foundation

Shapin, Steven, and Simon Schaffer. 1985. *Leviathan and the Air-Pump: Hobbes, Boyle, and the Experimental Life.* Princeton: Princeton University Press.

Shiva, Vandana. 2000. Tomorrow's Biodiversity. New York: Thames and Hudson.

Sievanen, Leila, Brian Crawford, Richard Pollnac, and Celia Lowe. 2005. "Weeding Through Assumptions of Livelihood Approaches in ICM: *Eucheuma* Farming in the Philippines and Indonesia." *Ocean and Coastal Management* 48: 297–313.

Sody, H.J.V. 1949. "Notes on Some Primates, Carnivora, and the Babirusa from the Indo-Malayan and Indo-Australian Regions." *Treubia: Journal of Zoology, Hydrobiology, and Oceanography of the Indo-Australian Archipelago*. Buitenzorg (Bogor): Koninklijke Plantentuin Van Indnonesie. 20 (Part 2):121–43.

Soekarno. 1989. *Coral Reef and Associated Habitats in the Vicinity of Bunaken, Togian Islands, and Takabone Atoll: Their Conservation Value and Needs*. Prepared for USAID Natural Resources Management Project. Jakarta: Center for Oceanological Research and Development.

Sopandi. 1997. *Jelajah Etnik: A Journey Through Wallacea*. Privately published museum catalogue.

Sopher, David E. 1977 (1965). *The Sea Nomads: A Study of the Maritime Boat People of Southeast Asia*. Singapore: National Museum of Singapore.

Soulé Michael. 1995. *Reinventing Nature: Responses to Postmodern Deconstruction*. Washington: Island Press.

Spyer, Patricia. 2000. *The Memory of Trade: Modernity's Entanglements on an Eastern Indonesian Island*. Durham: Duke University Press.

Spivak, Gayatri Chakravorty. 1998. *A Critique of Postcolonial Reason: Toward a History of the Vanishing Present*. Cambridge, MA: Harvard University Press.

Steedly, Mary. 1993. *Hanging Without a Rope: Narrative Experience in Colonial and Post-colonial Karoland*. Princeton: Princeton University Press.

Stewart, Kathleen. 1996. *A Space on the Side of the Road: Cultural Poetics in an "Other" America*. Princeton: Princeton University Press.

Stoler, Ann. 1995. *Race and the Education of Desire: Foucault's History of Sexuality and the Colonial Order of Things*. Durham: Duke University Press.

———. 2002. *Canal Knowledge and Imperial Power: Race and the Intimate in Colonial Rule*. Berkeley: University of California Press.

Sturgeon, Janet C. 2005. *Border Landscapes: The Politics of Akha Land Use in China and Thailand*. Seattle: University of Washington Press.

Sundjaya. 2005. The Community Base Marine Conservation Togean Islands. *http://www.conservation.or.id/site/modules/detail.daily.php?textid=4387633321928321* (accessed 7/29/05).

Supriadi, Dedy, and S. Akbar. 1997. "Togean Macaques: The New Species of Sulawesi Macaques is Threatened." *Tangkasi* 3(1&2):10–11.

Supriatna, Jatna. 1991. "Hybridization Between *Macaca maurus* and *M. tonkeana*: A Test of Species Status Using Behavioral and Morphogenetic Analyses." Ph.D. Dissertation. Department of Anthropology. Albuquerque: University of New Mexico.

Surjadi, Purbasari, and Jatna Supriatna. 1998. "Bridging Community Needs and Government Planning in the Togean Islands, Central Sulawesi, Indonesia." Proceedings of the International Tropical Marine Ecosystems Management Symposium. Port Townsend: Great Barrier Reef Marine Park Authority.

Tagliacozzo, Eric. 2005. "The Lit Archipelago: Coast Lighting and the Imperial Optic in Insular Southeast Asia (1860–1910)." *Technology and Culture* 46(2):306–28.

Tempo. 1999. Orang Bajau, Dalam Untung dan Malang. March 15, 1999.

Terborgh, John. 1999. *Requiem for Nature*. Washington D.C.: Island Press.

Toer, Pramoedya Ananta. 1997. *The Mutes Soliliquy* [*Nyanyi Sunyi Seorang Bisu*]. Jakarta: Lentera.

———. 1996. *This Earth of Mankind* [*Bumi Manusia*]. Max Lane, trans. New York: Penguin Books.

Tronto, Joan. 1993. *Moral Boundaries: A Political Argument for an Ethic of Care*. New York: Routledge.

Tsing, Anna. 2004. *Friction: An Ethnography of Global Connection*. Princeton: Princeton University Press.

———. 1993. *In the Realm of the Diamond Queen: Marginality in an Out-of-the-Way Place*. Princeton: Princeton University Press.

Umbgrove, J.H.F. 1930. "Madreporaria from the Togian Reefs" (Gulf of Tomini, North-Celebes).

———. 1939. "De Atollen en Barriere-Riffen der Togian-Eilanden." *Leidsche Geologische Mededeelingen*. 11:132–87.

Volkman, Toby Alice. 1990. "Visions and Revisions: Toraja Culture and the Tourist Gaze." *American Ethnologist* 17(1):91–108.

van der Wal. 1680. "Extract uit het Journaal van den Hoeker de Brandgans . . ." In *Reistogten in de Afdeeling Gorontalo, Gedaan op Last der Nederlandsch Indische Regering*. Amsterdam: F. Muller. C.B.H von Rosenburg, 1865.

van Musschenbroek, S.C.J.W. 1880. "Toelichtingen Behoorende Bij de Kaart van de Bocht van Tomini of Gorontalo en Aangrenzende Landen, de Reeden, Afvoerplaatsen, Binnenlandsche Wegen en Andere Middelen Van Gemeenschap" (Explanation Accompanying the Map of the Bay of Tomini or Gorontalo and Adjacent Lands, Anchorages, Ports, Local Roads, and Other Means of Communication). *Tijdschrift van het Aardrijkskundig Genootschap, Vierde Deel*. Amsterdam: C. L. Brinkman.

Walden, Arthur. 1871. "On the Birds of the Isles of Celebes." *Proceedings of the Zoological Society of London*. 22: 329–37.

———. 1872. "On Some New Species of Birds from Celebes and the Togian Islands." *Annals of the Magazine of Natural History* 4(9):398–401.

Walhi. 2003a. *Bebaskan Nelayan Dari Penindasan TNC* (The Nature Conservancy) *di Perairan Komodo* (Free Fishers from the Oppression of TNC in Komodo Waters). Jakarta: Walhi.

———. 2003b. *Tim Advocacy Masyarakat Komodo* (Komodo Peoples Advocacy Team). Jakarta: Walhi.

Wallace, Alfred Russel. 1863. "On the Physical Geography of the Malay Archipelago." *Journal of the Royal Geographical Society*. 33:217–34.

Warren, Carol, and Kylie Elston. 1994. *Environmental Regulations in Indonesia*. Nedlands: University of Western Australia Press.

Warren, James. 1981. *The Sulu Zone: 1768–1898*. Singapore: Singapore University Press.

Weber, Max. 1946[1914]. *From Max Weber.* Gerth and Mills, eds. New York: Oxford University Press.

Werner, Carol Thuman, and Jane Maxwell. 1992. *Where There is no Doctor.* Hesperian Foundation.

Whalley, Christine. 2004. *Rough Waters: Nature and Development in an East African Marine Park*. Princeton: Princeton University Press.

Whitten, Anthony, Muslimin Mustafa, Gregory Henderson. 1987. *The Ecology of Sulawesi*. Yogyakarta: Gadjah Mada University Press.

Wilson, Edward O. 1992. *The Diversity of Life*. New York: Norton.

———. 1986. *Biodiversity.* Washington: National Academy Press.

World Rainforest Movement. 2004. "The Nature Conservancy's Plans in Komodo National Park." WRM Bulletin No. 80. Montevideo: World Rainforest Movement.

Yuliati, Susi, Shaifuddin Akbar, Yakup Hutabarat, Chalid Mohammad, Salahuddin. 1994. *Laporan Studi Sistem Pengetahuan dan Teknologi Pemanfaatan Tumbuhan dan Hasil Laut: Penelitian Etnobiologi Kepulauan Togian* (Report on the Study of Knowledge Systems and the Useful Technology for Using Plants and Sea Resources: Togean Island Ethnobiology). Depok: Yayasan Bina Sains Hayati Indonesia.

Zerner, Charles. 2003. "Sounding the Makassar Strait: The Poetics and Politics of an Indonesian Marine Environment." In *Culture and the Question of Rights: Forests, Coasts, and Seas in Southeast Asia*, Charles Zerner, ed. Durham: Duke University Press.

———. 1994. "Through a Green Lens: The Construction of Customary Environmental Law and Community in Indonesia's Maluku Islands." *Law and Society Review* 28(5): 1079–1121.

Index

The Shadows and Lights of Waco: Millennialism Today
by James D. Faubion

Life Exposed: Biological Citizens after Chernobyl by Adriana Petryna

Anthropos Today: Reflections on Modern Equipment by Paul Rabinow

When Nature Goes Public: The Making and Unmaking of Bioprospecting in Mexico by Cori Hayden

Picturing Personhood: Brain Scans and Biomedical Identity
by Joseph Dumit

Fiscal Disobedience: An Anthropology of Economic Regulation in Central Africa by Janet Roitman

Race to the Finish: Identity and Governance in an Age of Genomics
by Jenny Reardon

Everything Was Forever, Until It Was No More: The Last Soviet Generation by Alexei Yurchak

Wild Profusion: Biodiversity Conservation in an Indonesian Archipelago
by Celia Lowe

time for "countermapping"

What do these expanses of primary colors interspersed with rare habitations offer us as a site for understanding biodiversity and its conservation? The term "biodiversity" emerged as a new mode of biological and social organization in the United States in the mid-1980s.[4] Coming, as it did, after several decades of heightened attention to environmental risk, biodiversity, as a particular framing of nature and culture,[5] began to reorganize earlier notions of natural history, wilderness, taxonomy, ecology, natural variety, species, and the like. Biodiversity was not so much a solution to the problem of environmental risk, however, as its problematization. It instigated a new form of critical inquiry into the relationship between entities conceived of as "nature" and "the human." Thrust into the light was, on the one hand, nature, understood as the linkages between genetic variation, species populations, communities and ecosystems, and land and marinescapes and, on the other hand, humanity, with its ability to instigate what biologist Michael Soulé has termed the "sixth great extinction."

Simultaneously, biological science itself was restructured around the biodiversity problematic. The task of protecting and restoring biodiversity was articulated with the sciences of population genetics, evolutionary biology, systematics, landscape ecology, and the study of ecosystems to form the new field of conservation biology. Unlike nineteenth-century natural history, or twentieth-century wildlife biology, conservation biology is self-consciously "mission-oriented" and sees itself as comparable to medical research in its goal of intervening in ailing systems.[6] Conservation biology is unusual among the natural science disciplines in that its value orientation—identified in terms of biodiversity's utilitarian and inherent worth—is explicit. This new science sees its object of study as threatened, and describes the state of plants and animals in terms of crisis. As a scientific practice, it is focused on intervention and is self-consciously directed toward solving its urgencies.

Biodiversity also encompasses an important geographic dimension. Conceptualized through the variety and uniqueness of species, diverse life is not uniformly distributed. Rather, regions with large numbers of species, where many unique life forms are found, tend to be concentrated in the tropics. Conservation biologists recognize roughly twenty-five "hotspots" as having this hyperdiversity. Since most hotspots are located in the rainforests and on the coral reefs of the global South, the peoples of tropical nations—both non-EuroAmerican biologists, and those who live in close proximity to tropical flora and fauna—have taken on a particular significance within the biodiversity problematic.

Several elements have made this particular assemblage of nature and culture under the sign of biodiversity possible. First, biologists observe an exponential reduction in the diversity of species forms across the globe.

must travel along a path *between* the human and the wild profusion. This is the path we will follow here.

The Togean Islands and Biodiversity

The Togean Islands[2] a small archipelago in the middle of the eastward facing Gulf of Tomini, harbor a volcano, which erupted as recently as 1983, and six raised limestone islands.[3] Small, craggy, thinly soiled islets bordering the shores of the main islands create anchorages, mangrove-lined boat passages, and resource collecting sites for Togean Island peoples. Small settlements intermittently punctuate the shoreline; houses built from cement, wood, and other forest materials lie at the edges of the land, or on stilts over the fringing coral substrate. There are no telephones or newspapers, and the only road is in Wakai town on Batu Daka Island. Coconut palm and vegetable gardens spread from coasts into the interiors. Forests in the midst of these encroaching cultivations supply Togean people with canoe timber, sago palm, medicinal plants, and other useful vegetation. Togean forests are also home to many insects, herpefauna, and mammals of interest to biologists, who are concerned by evidence of forest clearing. Upon first glance, many signs of habitation in the landscape are hidden, however. One tends to notice only the overwhelming verdancy.

Surrounding Togean waters reflecting a violent equatorial sun contain coral reefs, sand banks, sea grass beds, and azure depths. Togean people collect subsistence and market-oriented marine goods in these waters, of which fish and sea cucumber (*trepang*, S:*bale**) are the most important. Beyond the reef, in deeper waters, pelagic fish school, drawing local fishers and commercial boats from the mainlands of North and Central Sulawesi. Ferry boats make irregularly scheduled rounds between the islands and the mainland towns of Gorantalo, Poso, and Ampana. To the south of the islands, the mountains of Central Sulawesi are visible. To the north, only the waters of the Gulf of Tomini are in view. Biologists are concerned with the health of Togean reefs and waters. People have fished the surrounding reefs with both dynamite and cyanide, and several kinds of sea creatures, like the Napoleon wrasse fish, are threatened with local extirpation. But when one looks out at the expanse of Togean waters, coral reefs, ferry routes, and fishing sites are obscured. One notices, at first, only various shades of blue.

* "S" represents a Sama language term throughout the book. The other terms in parentheses are Bahasa Indonesia, the Indonesian national language.

let them breathe, although later they would end up in a formaldehyde bath, and Budi would send them traveling to the Smithsonian Institution for confirmation of their uniqueness. Laboratories in Washington, D.C., and other EuroAmerican scientific institutions were important for determining the specificity of this place and for confirming Yakup and Budi's acts of discovery and nature-making.

We passed through Budi's plots marked off with strings, where he had kept track of three thousand trees of various sizes for the past several years. Through his marking, measuring, and counting, he was developing a scientific record of the trees in the forest. He knew their species names, when they would flower, and how fast they were growing. Further on down the trail, we descended into a cave that Yakup had found. The ceiling was lined with hanging bats awakened by our movements, and the floor was layered in guano. Yakup reminded me that it is not only human agency that is capable of transforming a place—bats are important for pollinating the trees of the Malenge forest. When we continued on Budi noticed a striped spider, ten centimeters long, hanging in its web between some leaves and vines. He took a photograph, a close-up still life that would later trigger memories of the walk, coding this place as nature and reminding us of its aesthetic perfection down to its smallest details.

In our movements through the Togean landscape, scientists' stories of species endemism vied with the narratives of plants and animals Togean people wanted to tell. Whereas biologists were most cognizant of the fig trees that provide food for the Togean macaque, an unusual monkey living on the island, Pak Ahmad was aware that the dipterocarps scientists value as signs of a "wild" forest are excellent trees for canoe-building. Walking along the path, Ahmad, who was born on Malenge Island, told us of snakes. Togean pythons have been known to eat deer, pigs, and even small children. Sliding his hand rapidly through the air, he showed us how a snake strikes. He and several others had once rescued a boy in a garden. A python had wrapped itself around the child and was beginning to take his breath away when they discovered him. They unwound the snake, tail first, before attacking it with their machetes.

Ahmad's ways of knowing Togean flora and fauna called biodiverse nature into question. Bees will pollinate and snakes will wind, yet what is deemed important in a landscape depends on who is looking. Biologists found monkeys and lizards intellectually gratifying, while Sama people found the monthly flowering of the sea grasses—an event overlooked by Togean biologists—to be aesthetically alluring. Natures are "made" at the intersection of humans with their particular social histories, and plants and animals with their unique evolutionary and ecological histories. Neither "science" nor "society" will tell us all the interesting things one might want to know about these natures. To proceed further, one

Introduction

BETWEEN THE HUMAN AND THE WILD PROFUSION

> [The naturalist] looks upon every species of animal and plant now living as the individual letters which go to make up one of the volumes of our earth's history; and, as a few lost letters may make a sentence unintelligible, so the extinction of the numerous forms of life which the progress of cultivation invariably entails will necessarily obscure this invaluable record of the past. It is, therefore, an important object, which governments and scientific institutions should immediately take steps to secure, that in all tropical countries colonized by Europeans the most perfect collections possible in every branch of natural history should be made and deposited in national museums, where they may be available for study and interpretation.
>
> —Alfred Russel Wallace, "On the Physical Geography of the Malay Archipelago"

In April of 1996, I hiked through the upland forest that runs along the narrow central crest of Malenge Island with two biologists from Jakarta, Yakup and Budi, and with Pak Ahmad, a local ranger hired by the scientists to work at their research station, Camp Uemata.[1] We were there to collect new forms of herpefauna: lizards, snakes, and frogs. Together we scrambled up muddy slopes, grabbing hold of verdant branches and shrubs to pull us up, breathing hard. We scanned the trees for pythons and the ground for lizards. In the clearings we stopped to gaze at the vistas of the coast, and we rested against the architecturally fabulous buttress roots of the forest's huge dipterocarps. When we began to be bitten by red ants, which never seemed to take very long, we would move again, eyes fixed on the trail and underbrush, poking through bogs with sticks, constantly on the lookout for tiny eyes peering back at us.

Species uniqueness and endemism were the salient features of place delimiting a Togean nature the biologists wanted to claim as "biodiverse." To do this they needed to observe, record, and document species that were only found in the Togean Islands. Yakup spotted some lizards with blue tails that he suspected were "new to science." We dove with our hands out—to the left and to the right of the trail—trying to grasp the elusive electric-blue tails. We placed the lizards in plastic bags with air holes to

Wild Profusion

ABBREVIATIONS

CI	Conservation International
DPR	Dewan Perwakilan Rakyat (National Parliament)
FFI	Fauna and Flora International
GIS	Geographic Information System
GOI	Government of Indonesia
IFABS	Indonesian Foundation for the Advancement of Biological Sciences
ITMEMS	International Tropical Marine Ecosystems Management Symposium
JICA	Japan International Cooperation Agency
LIPI	Lembaga Ilmu Pengetahuan Indonesia (Indonesian Institute of Science)
LSM	Lembaga Swadaya Masyarakat (People's Self-Improvement Leagues)
ODA	Overseas Development Agency
PKI	Partai Komunis Indonesia (The Communist Party of Indonesia)
TNC	The Nature Conservancy
UI	University of Indonesia
UNDP	United Nations Development Program
USAID	United States Agency for International Development
VOC	Verenigde Oostindische Compagnie (The Dutch East India Company)
Walhi	Wahana Lingkungan Hidup Indonesia (Indonesian Forum for the Environment)
WCS	Wildlife Conservation Society
WPTI	Wildlife Preservation Trust International
WWF	World Wide Fund for Nature

Seven women, in particular, have provided an exceptional level of friendship, confidence, and critical support that have been invaluable to my ability to thrive as a person and a scholar. I wish to express heartfelt appreciation to Suraya Afiff, Lyn Criddle, Miriam Kahn, Nancy Peluso, Lorna Rhodes, Laurie Sears, and Anna Tsing. Jane Bixby Weller, whose lionfish I first encountered on Ahé Island in the Tuamotu Archipelago, generously created the artwork, "Three Lines," for this book. Her image is a magical and much appreciated gift. Karl Taylor kindly granted me permission to use his image of Susunang village. Charles Zerner, who enjoyed my stories of the sea from the start, and Philip Yampolsky, with whom I traveled through the Sulawesi highlands in search of song, have also been inspiring. Thanks to Rick Mazzotta for his friendship and work on the manuscript in 2004.

Scholarship requires the faith and promissory notes of outside funders, and I have been fortunate in this regard. This research was supported by National Science Foundation Grant SBR-9628940, a Fulbright IIE under the sponsorship of the American-Indonesian Exchange Foundation, an S.V. Ciriacy-Wantrup Postdoctoral Fellowship in Natural Resource Studies from U.C. Berkeley, the University of Washington Graduate School, the University of Washington Center for Southeast Asian Studies, the Yale University Department of Anthropology, and a Mellon Fellowship in Humanistic Studies.

Earlier versions of three chapters of *Wild Profusion* have appeared in print before. Chapter 1 is revised from the article "Making the Monkey: How the Togean Macaque went from 'New Form' to 'Endemic Species' in Indonesians' Conservation Biology," in *Cultural Anthropology* 19(4):491–516 (© 2004 American Anthropological Association). Chapter 3 has been reworked from "The Magic of Place: Sama at Sea, on Land, in Sulawesi, Indonesia," in *Bijdragen tot de Taal, Land, en Volkenkunde* 159(1):109–133 (© 2003 KITLV Press). An earlier version of chapter 5 appeared as "Global Markets, Local Injustice in Southeast Asian Seas: The Live Fish Trade and Local Fishers in the Togean Islands of Sulawesi, Indonesia," in *People, Plants, and Justice: The Politics of Nature Conservation*, edited by Charles Zerner (© 2000 Columbia University Press, reprinted with permission of the publisher). At Princeton University Press I wish to thank my editor, Fred Appel, and two anonymous reviewers.

Finally, I wish to thank my family for their encouragement, and for supporting my adventures and sense of curiosity over the years. To my mother, Jeanette, for insisting that I write for a wide audience; to my father, Richard, for caring enough to ask if the book is done yet, and to my brother, Steven, for inspiring me through his own courage and success. I dedicate *Wild Profusion* with much admiration and love to them.

Harwell, Cathryn Houghton, Jake Kosek, Hugh Raffles, Curtis Renoe, Janet Sturgeon, and Eric Tagliacozzo each contributed to my thinking.

Two intellectual communities in California changed forever how I view my work. I was privileged to be a visiting scholar in anthropology at U.C. Santa Cruz in 1997–98. A particularly gracious and influential learning community, U.C. Santa Cruz transformed the way I view the possibilities for my discipline and for scholarship in general. Tim Choy, Melanie Dupuis, Donna Haraway, Cori Hayden, Sarah Jain, Ravi Rajan, Shiho Satsuka, and Jonathan Scheuer were supportive in this regard. There are simply not enough words to express my gratitude to Anna Tsing, who risked the ferry ride to the Togean Islands, and whose work on Kalimantan is my model for scholarly engagement with Indonesia. In 2002–3, as an S.V. Ciriacy-Wantrup fellow in Natural Resource Studies at U.C. Berkeley, I had the good fortune to work again with Nancy Peluso in the Department of Environmental Science, Policy, and Management. The Berkeley environmental politics community, including Suraya Afiff, Amita Baviskar, Lea Borkenhagen, Louise Fortmann, Gillian Hart, Donald Moore, and Michael Watts, were key to my rethinking of cultures of nature. In anthropology, Paul Rabinow was especially gracious to read my work and to help me develop my thinking in science studies. His writings have been especially influential.

Since I arrived at the University of Washington in 1999, I have found the Department of Anthropology to be a generous and supportive intellectual community that has encouraged my disciplinary thinking and teaching in a variety of ways. Ann Anagnost, Laada Bilaniuk, Rachel Chapman, James Green, Daniel Hoffman, Miriam Kahn, Charles Keyes, Peter Lape, Rebecca Lemov, Kathleen O'Connor, Arzoo Osanloo, Lorna Rhodes, Kalyanakrishnan Sivaramakrishnan, and Janelle Taylor have all inspired my work. The Center for Southeast Asian Studies at the University of Washington has also continually supported me intellectually and otherwise. Kiko Benitez, Mary Callahan, Patrick Christie, Judith Henchy, Christoph Giebel, Daniel Lev, Marjorie Muecke, Vince Rafael, Pauli Sandjaja, Laurie Sears, and Sara VanFleet have all been excellent colleagues and friends. Laurie Sears and Daniel Lev each generously read the entire manuscript and provided much appreciated feedback. I am grateful to Wolfgang Linser who assisted me in finding and interpreting Dutch language historical documents. Tani Barlow of the Critical Asian Studies Program, and Kathleen Woodward of the Simpson Humanities Center, have also supported my work. My students at University of Washington—Cheryll Alipio, Chris Brown, Hanh Bich Duong, Diane Fox, David Giles, Yu Huang, Pat McCormick, Chingchai Methaphat, Rebeca Rivera, Leila Sievanen, Mia Siscawati, Asep Suntana, Woonkyung Yeo—have been significant interlocutors.

ACKNOWLEDGMENTS

THE JOURNEY required to research and write a book such as this is always multisited, and entails a proliferation of debts of gratitude. First and foremost I wish to acknowledge the many Indonesian people who made this work possible. In the Togean Islands, scientists and Sama people brought me along and taught me what they knew of everything from watching birds to diving for sea cucumber, from counting butterfly fish to cyanide fishing, often at risk to themselves. The conventions of anthropology and the nature of my work dictate that I am obliged to rely on pseudonyms in this book. Except in the case of public figures and scientists with extensive publishing records, the personalities in *Wild Profusion* are composite characters, although the events and statements all actually happened or were related to me. Disguised as they may be, this book is written with very real gratitude to those actual persons who stand behind the figures I write of here.

In Indonesia, I am grateful to the Indonesian Institute of Science (Lembaga Ilmu Pengetahuan Indonesia) for granting permission to conduct this study. Nelly Paliama of the Jakarta Fulbright office deserves special mention for her dedicated friendship and tireless assistance. I especially wish to thank the many friends and colleagues from the Indonesian conservation community who helped along the way: Suraya Afiff, Pak Amir, Susi Darnedi, Rili Djohani, Christo Hutabarat, Reed Merrill, Mochamad Indrawan, Tim Jessup, Reny Juita, Adrian Lapian, Pak Pili, A. Hadi Pramono, Ida Purbasari, Chris Rotinsulu, Ari Supandi, Dedy Supriady, Jatna Supriatna, Wendy Tan, Fadjar Thufail, Graham Usher, and Arif Wicaksono. The Wewengkang-Tengker family gave me a home I could always return to in Manado and I am forever appreciative of their kindness. Several Indonesian and international conservation NGOs also provided invaluable assistance. I wish to thank Conservation International, Kelola, The Nature Conservancy, Walhi, the World Wide Fund for Nature, Yayasan Bina Sains Hayati Indonesia, Yayasan Ibnu Chaldun, and Yayasan Sejati.

This work began as a dissertation project in the Department of Anthropology at Yale University where Harold Conklin, Nancy Peluso, James Scott, and Tinuk Yampolsky each contributed the selfless act of teaching, and where Joseph Errington provided invaluable guidance as dissertation advisor. I was fortunate to have an amazing cohort of fellow student colleagues each working on topics of nature and/or Southeast Asia at Yale: Gene Ammarell, Amity Doolittle, Chris Duncan, Lisa Fernandez, Emily

particular problems of nature and nation within the degrees of freedom and constraint encompassed by late–New Order Indonesian conditions. As an analytic, reason allows us to move beyond good and evil, in Nietzsche's sense, to construct a genealogical understanding of biodiversity conservation in Indonesia.

In *Wild Profusion* I describe Indonesians' science and Togean peoples' natures in the context of the structures of subjection and abjection within which the Indonesian people were living in the waning years of the New Order. Rather than a story of the horrors of the time, however, my intention is to provide here a more hopeful narrative. In reflecting upon the possibilities and limits of 1990s biodiversity conservation in Indonesia through an analytic of postcolonial science, and through genealogies of reason, I desire to introduce new possibilities for thought and to open new spaces in which Indonesian scientists, Togean people, and Sulawesi natures each will be able to live, even to thrive.

Monetary Fund and the world financial community withdrew its support from Indonesia.

As a consequence, the Indonesian people began to raise their voices in a manner unheard of since the early 1960s under Indonesia's first president, Sukarno. Protests and riots broke out in cities across the country. Cries for "reformation" (*reformasi*) were met with some (temporary) restraint by the Army, which was not in utter disagreement with the popular perception that things were going poorly. When three students were shot in a protest at Trisakti University in Jakarta in May of 1998, that was the end for Suharto, Asia's longest reigning dictator and leader of a state best characterized by crony capitalism, coercive governance, and its leader's "darling status" amongst northern governments during the Cold War. The end of the Suharto regime led to social upheaval, including the rapes and murders of Chinese Indonesians in Jakarta, Muslim-Christian violence across the Maluku Province and Dayak-Madurese fighting in Kalimantan, a series of shopping mall arsons in major cities, the re-emergence of "ninja" killings in Central and East Java and, most notably, a referendum granting independence to East Timor (occupied by Indonesia since 1975).

These many sharp transitions unsettled how so many of us had been thinking about Indonesia. Although everyone understood that Suharto could not live forever, it seemed impossible, after thirty-three years, to imagine an Indonesia without him. Due to the strength and duration of his rule, the man had become a metonym for the state and all of its problems. It had been easy to place the Indonesian nation within the black box of the "Suharto" state, and to blame everything on the man himself—from coercive national language policies, media censorship, state rituals domesticating cultural and ethnic difference, to unequal natural resource policies. The effect Suharto's fall from power had on many of us, foreign scholar and Indonesian citizen alike, was that for the first time we were forced to think beyond good and evil. If Indonesians were going to invent for themselves a new nation-state in the post-Suharto era, they could only do so based on their experiences of life during those darker times.

In relation to Togean science, this suggests I needed to search for the counterhegemonic forms of thought and action that might have existed previously *within* neoliberal, state-supported, biodiversity conservation. Those moments, where Indonesian scientists were thinking thorough the possibilities of the nation-form from within the constraints of the New Order state and transnational capital, were one location for the alternative political imaginations that might now become transformative or even libratory in the post-Suharto era. Along the way, an analytic of thought, or "reason," has become useful to my project. Reason allows me to put Sama and scientist into the same frame, each group struggling to solve

"in action," and to understand the co-constitution of science and society, things and people. Postcolonial perspectives, alternatively, have provided a framework for understanding identity and subjectivity; postcolonial theory develops the sense of politics missing from much scholarship in science and technology studies. *Wild Profusion* is, thus, situated within the emerging theoretical terrain of postcolonial science studies in an attempt to hold in tension the diverse ways of seeing, knowing, and being that belong to both Indonesian scientists and Sama people.

An additional consequence of the mid-1990s political economy approach to anthropology was that it was easy to lose sight of "nature" and the fact of documented losses in the abundance and variety of life forms on earth. While biologists have often been misguided in their methods for convincing the rest of us of the value of biodiverse nature, this says nothing of the existence (or disappearance) of plants and animals themselves. As Vandana Shiva (2000) has demonstrated so clearly, to be against coercive neoliberal conservation or the corporatization of life is not to be against life itself in a plurality of forms. It is, in fact, to recognize the widest possible diversity of life—especially those forms of agricultural diversity that conservationists tend to overlook, and those forms of cultural diversity that are so easily pushed aside by state, corporate, and scientific natures.

Another issue for the story of *Wild Profusion*, then, is how to maintain an antifoundationalist perspective on nature while simultaneously remaining attentive to the stories of nature biologists wish to tell. It would make no sense, after all, to be in favor of *less* biological diversity. Nor does it make sense to see biologists and conservationists as the most harmful actors in scenarios of social abjection or schemes to remake nature in one's own image. This will not be a book written against the diversity of life or against the biodiverse natures biologists make. It is, however, a narrative that questions the methods we use to understand and maintain nature's variety, and that asks whose vision of nature the wild profusion can and should sustain.

In retrospect, the fires were only one kind of conflagration to sweep across Indonesia in 1997–98. When I left the Togean Islands that September, making my way back to the city of Manado to depart from Indonesia, I discovered that the rupiah, the Indonesian currency, had fallen precipitously. This was the start of the Asian economic crisis, which began with currency scandals in Thailand and ended with currency devaluation and widespread economic dislocation across the region. As the value of the rupiah continued to drop, prices for basic goods and necessities (*sembako*), such as cooking oil, kerosene, rice, sugar, tea, and chili pepper, began to rise to untenable levels. At the same time, the International

instrumental rationality for others' natures and ideas; second, see how these regimes can extend the power of the state through coercive practices of cultural and economic mystification; and third, to recognize the ways economic development programs continually fail to improve human welfare while conservation initiatives are hardly ever able to save nature.

Once I was in Indonesia, however, this theoretical background placed me in a quandary. While at first it was difficult to see anything else in the Togean project but political and economic injustice in many forms, I was also developing a certain friendship with, and anthropological loyalty to, the scientists and activists whose work I was following. These scientists were deeply committed to their program of conservation and development, and saw their work as a point of hope for the nation. Moreover, one of the very first things I discovered in Indonesia was that scientists were quite sensitive to criticism from foreign scholars and were continually evaluating me as to what kind of person I was: was I the type of foreign scholar who wanted to tell them how they had it all wrong, or could I also learn from them and value their contributions? How was I to maintain my analytic perspective on the political logics of conservation and development, and simultaneously continue to share an empathic relationship with Indonesian scientists who believed that conservation biology and projects of economic development would contribute to the advancement of their nation?

It was ultimately my Indonesian colleagues themselves who taught me to see their work in a new light. In their questioning of my understanding of conservation, and in their reading of my theories as criticism of them (as individuals and as a national community), it became clear that the problem of how to hold the situatedness of Indonesian's science in simultaneous tension with the questions of justice raised by Togean conservation was central to my work and should be explored rather than delicately avoided. I would need to find a way to take Indonesian scientists and their Togean work seriously, while retaining my awareness that biodiversity conservation was capable of further marginalizing Togean people and their ability to live within the natures and histories they themselves sustained.

My desire simultaneously to understand scientists' and Sama peoples' perspectives on Togean natures, even when they were in conflict, led me down an empirical path marked by an increased attention to the details of biology and conservation I witnessed, and by noticing differences not only between, but *within*, the categories "Sama" and "Indonesian scientist." In the realm of theory, this practice led me to new literatures in science and technology studies and in postcolonial theory. The ethnomethodological approach of recent science and technology studies has helped me to track the micropractices of both scientists and Sama people

of concern for social and environmental activists, scientists, and for many fishers and farmers across the country who were suffering from both degraded environments and from accusations that they were responsible for destroying nature. My narrative approaches the status of Indonesian land and marinescapes through one specific framing of nature: biodiversity. Biodiversity, in 1990s Indonesia, implied a particular assemblage of nature, nation, science, and identity—making Indonesians' mapping of the concept distinct within the transnational biodiversity practices and discourses circulating widely at the time. As with the fires, Indonesia's greatest environmental disaster, biodiversity loss entails complex struggles over responsibility and blame, finance and profit, technoscience, nationalism, cultural and ethnic diversity—even over the nature of nature itself.

In *Wild Profusion* I tell of Indonesian scientists and their efforts to document the species of the Togean Islands, to economically and socially "develop" Togean people, and to turn the archipelago into a new national park. These elite Indonesians sought to produce a biological science that was both internationally recognizable and politically effective in the Indonesian context in order to incite positive social and environmental outcomes.

At the same time, I describe experiences of Togean people, especially people of Sama ethnicity. Sama and other Togean residents, as marginal Indonesian citizens, were often blamed for destroying the environment, despite the fact that well-connected entrepreneurs and state development projects were responsible for the major transformations in Togean land and marinescapes. Togean people had their own understandings of nature and nation, however, and stories of nature-making can also be narrated from their perspectives.

Wild Profusion is likewise a story of Togean nature. It details changes in Togean land and marinescapes, and contests over what should count as significant or valuable forms of nature for Togean peoples, traders and bureaucrats, tourists, and international and domestic biologists. It tracks nature's ability to shift its "nature" depending on who is doing the observing. And it takes seriously the decline in species diversity and abundance on a global scale.

I began to study the intersection of Indonesians' conservation biology, Sama lives, and Togean nature in 1993. My initial approach to this convergence was shaped by my background in political ecology and in the critical perspectives of what anthropologists were calling then a "culture and political economy" school. In the 1990s such an approach provided a clear formula for determining winners and losers in the transnational political economy. Such analyses of both environmentalism and economic development prepared me well to: first, recognize how conservation and development regimes have tended to substitute a belief in pure nature and

PREFACE

AS I DEPARTED for the last time from the Togean Islands of Central Sulawesi, in August of 1997, the sun was blood red and the horizon enveloped by a smoky haze. Indonesia was burning. It was the dry season, but this would be like no dry season ever before. Over the ensuing months, the dense haze would form a cloud of smoke across Southeast Asia as wide and broad as the footprint of the United States. Fires of that El Niño year would cause planes to crash and ships to run aground, harm public health, destroy property and livelihoods, deplete biodiversity, and force President Suharto to issue a public apology to Indonesia's Southeast Asian neighbors. This dry season would prove to be an unprecedented environmental disaster for Indonesia.

Many different explanations for the fires emerged (Colfer 2002; Harwell 2000). For the most part, the state pursued its conventional strategy of blaming swidden agriculturalists and subsistence farmers for degrading the land and igniting the flames. Sometimes it was suggested that the sparks necessary to start the fires were caused by trees rubbing together or by lightening bolts. Authorities declared that anyone who attributed a cause to the fires other than the unusual El Niño was a "Communist," which effectively prevented the media from speculating on alternative explanations. Some nationalists wanted to blame foreign corporations, despite the fact that foreign investors in the Indonesian forestry sector all had domestic allies, including President Suharto's own family and business partners.

But an unprecedented triumph also emerged out of the smoke and ash: Indonesian nongovernmental organizations (NGOs) were able to counter the state's rhetoric that blamed the devastation on small-scale farmers, the rubbing of tree-limbs, foreigners, and the weather. For the first time, activists were able to make the case that the fires were coming from state-owned forests and oil palm plantations and were a consequence of the government's own forest conversion policies. They made this argument by appropriating the new scientific technology of Geographic Information Systems (GIS), pinpointing the exact locations of the fires and mapping these locations onto state and commercial properties. Produced through "science," the satellite images were politically unassailable, and the Indonesian Forum for the Environment (Walhi) was able to use them to bring lawsuits against logging corporations despite the status and power of the major actors involved in forest conversion and timber harvesting.

Wild Profusion tells the story of a time (the decade culminating in the fires of 1997–98) when the Indonesian environment had become a point

CONTENTS

Published by Princeton University Press, 41 William Street, Princeton, New Jersey 08540
In the United Kingdom: Princeton University Press, Oxford

Library of Congress Cataloging-in-Publication Data

Lowe, Celia, 1961–
Wild profusion : biodiversity conservation in an Indonesian archipelago / Celia Lowe.
p. cm.—(In-formation series)
Includes bibliographical references (p.)
ISBN-13: 978-0-691-12461-2 (cl : alk. paper)
ISBN-10: 0-691-12461-2 (cl : alk. paper)
ISBN-13: 978-0-691-12462-9 (pbk.)
ISBN-10: 0-691-12462-0 (pbk.)
1. Biological diversity conservation—Indonesia—Togian Islands. 2. Biological diversity conservation—Indonesia—Togian Islands—Social aspects. I. Title. II.In-formation series (Princeton University).

QH77.I5L69 2006
333.95′1609598—dc22 2006041598

British Library Cataloging-in-Publication Data is available

This book has been composed in Sabon

Printed on acid-free paper. ∞

pup.princeton.edu

Printed in the United States of America
1 3 5 7 9 10 8 6 4 2

Wild Profusion

BIODIVERSITY CONSERVATION IN AN INDONESIAN ARCHIPELAGO

Celia Lowe

PRINCETON UNIVERSITY PRESS
PRINCETON AND OXFORD

Series Editor
Paul Rabinow

A list of titles in the series appears at the back of the book

Wild Profusion